符号中国 SIGNS OF CHINA

中国玉

CHINESE JADES

"符号中国"编写组 ◎ 编著

中央民族大学出版社
China Minzu University Press

图书在版编目(CIP)数据

中国玉：汉文、英文 /"符号中国"编写组编著. —北京：
中央民族大学出版社，2024.3
（符号中国）
ISBN 978-7-5660-2295-0

Ⅰ.①中… Ⅱ.①符… Ⅲ.①玉石-介绍-中国-汉、英 Ⅳ.①TS933.21

中国国家版本馆CIP数据核字（2023）第255795号

符号中国：中国玉 CHINESE JADES

编　　著	"符号中国"编写组
策划编辑	沙　平
责任编辑	舒　松
英文指导	李瑞清
英文编辑	邱　械
美术编辑	曹　娜　郑亚超　洪　涛
出版发行	中央民族大学出版社
	北京市海淀区中关村南大街27号　　邮编：100081
	电话：（010）68472815（发行部）　传真：（010）68933757（发行部）
	（010）68932218（总编室）　　　　（010）68932447（办公室）
经销者	全国各地新华书店
印刷厂	北京兴星伟业印刷有限公司
开　　本	787 mm×1092 mm　1/16　印张：12
字　　数	152千字
版　　次	2024年3月第1版　2024年3月第1次印刷
书　　号	ISBN 978-7-5660-2295-0
定　　价	58.00元

版权所有　侵权必究

"符号中国"丛书编委会

唐兰东　巴哈提　杨国华　孟靖朝　赵秀琴

本册编写者

姜莉君

前言 Preface

玉在中国人的心目中，是美的化身，寓意纯洁、美好和高贵。中国人自古就崇玉、尚玉、佩玉、赏玉、藏玉。

早在七八千年前的远古时代，华夏祖先就发现并开始使用玉，当时人们佩戴用玉制成的装饰品，也将玉加工成简单的工具。玉在中国人的生活中扮演着重要的角色。在古代，玉是王权、礼制的信物，还是美好品德的象征，并被认为是吉祥之物，是人们对美好生活的诉求寄托。现在，人们使用精美的玉雕作

Jade is the incarnation of beauty in the eyes of Chinese, which symbolizes purity, fineness and nobility. Therefore, it has long become a tradition for Chinese to exalt jade to a high position through worship, adornment, appreciation and collection.

Far back to seven or eight thousand years ago, Chinese ancestors discovered jade and began to wear it as an ornament or make it into basic tools for use. From then on, jade has played a significant role in the life of Chinese. In ancient times, jade was the token of majesty and ritual norms (ritual norms refers to *Li*, a concept celebrated in Chinese Confucian philosophy to emphasize the significance of social norms and stratum), and also the emblem of auspice to which people entrusted their aspiration for happy life. Nowadays, people use the exquisite jade sculpture for display and ornament, and as well purchase them for

为陈设品，佩戴用玉制成的装饰品，还购买玉器用作收藏或投资。可以说，玉在中国是用途最广、历史最长并独具魅力的一种材料，几乎没有哪一种材料像玉这样长期受到中国人的喜爱与推崇。

本书以中英文对照的形式，向海内外读者介绍了中国玉的历史与文化、玉石之美以及选玉技巧等方面的内容，配以大量精美的古今玉器图片，旨在让读者在买玉、赏玉、佩玉之余更深刻地领略中国玉的独特魅力。

private collection or investment. As it were, jade is the most widely-used, long-standing and fascinating material in China, while few stuff can enjoy such a long-term preference and esteem.

This Chinese-and-English compared book is dedicated to introducing the history and culture of Chinese jade, as well as its beauty and selection skills, to the readers both at home and abroad. Illustrated with a number of beautiful ancient and modern pictures about jade wares, it's aimed at impelling readers to attain a deep understanding about the distinctive glamour of Chinese jade while purchasing, appreciating and wearing it.

目录 Contents

中国有美玉
Beautiful Jade in China 001

从女娲补天的传说说起
From the Chinese Myth: Nvwa
　　　Mending the Heaven 002

玉的象征
The Symbol of Jade 043

玉之美
The Beauty of Jade 057

名玉之美
The Beauty of Famous Jade 058

造型之美
The Beauty of Shape 081

纹饰之美
The Beauty of Pattern 128

选玉技巧
The Skills to Choose a Jade Article 145

如何选购古玉
How to Buy an Antique Jade Article 146

如何选购现代玉
How to Buy a Modern Jade Article 163

附录
Appendix 175

玉器盘玩注意事项
Some Tips on Jades Treatment 175

玉器保养注意事项
Tips on Jades Preservation 178

中国有美玉
Beautiful Jade in China

中国盛产美玉,有玉石王国之称。中国的玉历史源远流长,距今已有七千多年。在这绵延千百年的崇玉、尚玉、佩玉、藏玉、赏玉的风尚中,中国人赋予了玉以灵性、礼教、德性等精神内涵,使玉成为中国传统文化中的重要内容。

Known as "Jade-Kingdom", China produces jade in great abundance with a long history which can be traced back to seven thousand years ago. In this millennia prevailed custom of worship, adornment, collection and appreciation of jade, Chinese endowed it with several spiritual connotations of intelligence, ritual and ethics, which make the jade an important content of traditional Chinese culture.

> 从女娲补天的传说说起

传说在远古时期,颛顼(zhuānxū)与共工为争帝位而争斗,共工被打败后一怒之下撞向不周山。不周山就是现在的昆仑山,是天地之间的支柱。不周山被共工撞断后,天地之间发生了巨变——天空向西北方向倾斜,大地向东南方向塌陷,洪水泛滥,大火蔓延,黎民百姓生活在水深火热之中。女娲不忍子民们受灾难之苦,决心炼石以补苍天。她采集五色土为料,炼

> From the Chinese Myth: Nvwa Mending the Heaven

As the legend goes, far back to the ancient times Zhuanxu and Gonggong were fighting for the throne. Defeated, Gonggong smashed his head in a fit of anger against Mount Buzhou—part of today's Kunlun Mountains, a pillar holding up the heaven. Then immediately the Mount Buzhou collapsed and caused immense change between heaven and earth: the sky tilting towards the northwest and the earth shifting to the southeast, together with vast floods and

玉琮(良渚文化)
Jade *Cong* (Liangzhu Culture, 4000-5000 years ago)

- 女娲补天
  Nvwa Mending the Heaven

出五色巨石来补天。天补好后，还需要支撑东西南北四极的柱子，于是女娲将背负天台山的神鳌的四足砍下来支撑四极。而多余的补天石散落在大地上，就成为今天的美玉。

当然，传说终归是传说，这只是人们对未知事物的想象和引申。但是从这美丽的传说来看，先人们早就认识到玉是从石演变而来的。在远古时期，人们在制作、使用石制工具时，发现了比一般石头坚硬的玉石，就用它来制作成特殊工具；又因玉石具有与众不同的光泽

huge fires, turning the man's world into a living inferno. Seeing human suffering in great misery, the goddess Nvwa was quite distressed and resolved to collect five-color earth to refine huge five-color stones to patch up the heaven. And then she cut off four legs of a giant tortoise — who used to carry Mount Tiantai in the water — and used them to supplant the fallen pillars to support the four poles of the heaven. The rest of the stones were left unemployed, thus scattered on the earth and later became beautiful jade we see today.

Legend as it is, a creative imagination on the part of human beings regarding the unknown, yet it tells that the predecessors had already realized that jade comes from stone. In ancient times, when people made and used stone tools, they found jade, a kind of stone harder than the general, and managed it into special tools or non-instrumental ornaments given its unique luster and brilliant color. Therefore it can be concluded that the development of jade wares is an evolution and continuation of stone wares, and moreover it is based on the stone culture where the jade culture grows and thrives.

和色彩，便还用它做成装饰品等非工具物品。因此可以说，玉器是由石器演进而来的，是石器的继续与发展，而玉文化则是石文化的延续。

玉器的历史从新石器时代开始，自此，玉的使用在中国几乎没有间断

The history of jade wares begins from the Neolithic Age (4500-8500 years ago) and then jade has been constantly used in China ever since. So the significant and profound Chinese jade culture has experienced a long-term development.

背景知识

颛顼：相传是中华民族始祖黄帝之孙，部落的首领。

共工：相传为中华民族始祖炎帝的后裔，是炎帝分支共工氏部落首领。传说身为天神的他是人首蛇身，满头赤发，性情暴躁。

女娲：中国神话传说中的一位创世女神。关于她的传说除了"女娲补天"之外，还有"女娲造人"，传说人类是她用黄土仿造自己捏成的。

Background Knowledge

Zhuanxu: the legendary grandson of the Emperor Huang who is said to be the ancestor of Huaxia Chinese nation, also the chief of the tribe.

Gonggong: the legendary descendent of the Emperor Yan who is also said to be the ancestor of Chinese nation, also the chief of the Gonggong tribe. In Chinese mythology Gonggong is an ill-tempered god with a human head, a serpent body and red hairs.

Nvwa: a legendary goddess and creator who mended the heaven and created human beings out of yellow clay in her own image.

● 颛顼
Zhuanxu

过。博大精深的中国玉文化，经历了漫长的发展历程。

新石器时代玉器

在距今约七八千年前的新石器时代早、中期，中国玉器进入初创阶段。1982年考古人员在位于内蒙古的兴隆洼文化遗址中发现的一对精美的玉玦，是目前所知的世界范围内最早的玉器。而在距今四五千年的新石器时代晚期，中国玉器制作的第一个高峰期出现了。

Jades of Neolithic Age

At the earlier and middle period of the Neolithic Age (4500-8500 years ago), that is, seven or eight thousand years ago, Chinese jade wares came into being. The pair of exquisite jade *Jue* (ritual object in shape of gapped disc) excavated by archaeologists in 1982 at Xinglongwa site in Inner Mongolia is known as the earliest jade artifact so far in the world. Three thousand years later, the late period of the Neolithic Age witnessed the first summit of jade development.

In the Neolithic Age (4500-8500 years ago), jade wares were abundant in variety and quantity with raw material such as jade and color stones like metamorphic marble-

- 玉刀（新石器时代）
Jade Knife (Neolithic Age, 4500-8500 years ago)

- 玉铲（新石器时代）
Jade Spade (Neolithic Age, 4500-8500 years ago)

新石器时代，玉器品种多，数量也多，其原料既有玉，也有漂亮的石头，例如与变质大理石矿共生的透闪石原矿等，但现在人们都把它们统称为"玉器"。这些玉器有从石器中分化出来的玉刀、玉斧、玉铲等玉工具，但数量更多的是用作祭祀的玉礼器，如太湖流域良渚文化时期（距今4000～5000年）的玉琮、三叉型器等。另外，还有一些作为部落图腾的象形玉器，如辽河流域红山文化时期（距今5000～6000年）的玉龙、玉猪等。

aggregated tremolite, which are generally called jade wares nowadays. These jade wares include jade tools derived from stone wares, such as jade knife, jade axe, jade spade, etc., and jade ritual objects which stood out in output for sacrificial ceremony like jade *Cong* and trident-shaped jade ritual wares made in the basin of Taihu Lake during the Liangzhu Culture Period (4000-5000 years ago). Besides, some pictographic jade wares were designed to serve as tribal totems like jade dragon and jade pig produced in the basin of Liao River during Hongshan Culture Period (5000-6000 years ago).

红山文化玉器

红山文化是中国东北地区的新石器文化，因最早发现于内蒙古自治区赤峰市红山后遗址而得名，在遗址处挖掘出土了大量的随葬玉器。

红山文化玉器以动物形与几何形佩饰居多。动物形佩饰以龟、鱼、鸟、龙等动物形象为主，多是人们信奉的灵物；几何形佩饰有方圆形玉璧、玉钺、勾云形玉饰、玉箍饰、双联或三联玉璧、玉环、玉珠和棒形玉等，多作为礼器使用。红山文化玉器多呈扁平状，图案抽象，线条简洁疏朗，多有穿孔，应为方便穿系佩戴。

Jade Wares of Hongshan Culture

Hongshan Culture (5000-6000 years ago) was a Neolithic culture in northeastern China and was named after Hongshanhou site at Chifeng City in Inner Mongolia where a number of jade funeral objects were excavated.

Jade wares of Hongshan Culture were primarily animal-shaped or geometric ornaments. The auspicious animals such as tortoise, fish, bird and dragon were often employed as decorative images, whereas geometric ornaments were mostly used to adorn ritual objects and varied greatly in shape. For instance, we have square-round jade *Bi* (round and plate-shaped jade ritual object with a square-round hole in the center), jade *Yue* (axe-shaped weapon, turning into ritual object afterwards), curved cloud-shaped jade ornament, jade hoop, jade *Bi* set (with two or three pieces combined together), jade ring, jade beads, stick-shaped jade, etc., most of which were flat in shape with abstract designs and simple outlines and string hole convenient for tying and wearing.

- 玉龙（红山文化）

红山文化的典型玉器，曾有"中华第一龙"的称号。

Jade Dragon (Hongshan Culture, 5000-6000 years ago)

Jade dragon is a typical jade ware of Hongshan Culture and once known as "The First Dragon of China".

- 玉兽形玦（红山文化）

红山文化的典型玉器，有人认为是龙，有人认为是猪，故又名"玉猪龙"。

Animal-shaped Jade *Jue* (Hongshan Culture, 5000-6000 years ago)

It is a representative jade ware in Hongshan Culture, half dragon-like and half pig-like, and therefore is also known as "jade pig dragon".

- 三联玉璧（红山文化）

Tri-cyclic Jade *Bi* (Hongshan Culture, 5000-6000 years ago)

- 玉发箍（红山文化）

Jade Hair-hoop (Hongshan Culture, 5000-6000 years ago)

良渚文化玉器

　　良渚文化是中国长江中下游太湖流域一支重要的古文明，因发现于浙江省余杭县（现余杭区）良渚镇而得名，以出土了大量的玉礼器而著名。

　　良渚文化玉器有琮、璧、钺、冠状器、三叉形器等器形。从功能上来看，大多是人类用来崇拜神灵的礼器。另外，还出现了以贯穿或缝缀等形式将各类玉饰件组合成的串饰，如串璜、管、珠、坠等。玉器上常见的纹饰有鸟纹、兽面纹、神人兽面纹、人面纹等，体现了史前人类对自然、图腾、祖先的崇拜。玉器造型、纹饰讲究对称，雕刻技法以阴刻为主，结合运用浅浮雕、圆雕、半圆雕、镂空等。

● 玉琮（良渚文化）
玉琮表面雕刻有神像飞鸟纹。
Jade *Cong* (Liangzhu Culture, 4000-5000 years ago)
The jade *Cong* has god figure and bird design carved on the surface.

● 玉璧（良渚文化）
Jade *Bi* (Liangzhu Culture, 4000-5000 years ago)

Jade Wares of Liangzhu Culture

Liangzhu Culture (4000-5000 years ago) constituted one important branch of ancient Chinese civilization developing around Taihu Lake in the middle and lower reaches of Yangtze River. It was initially discovered at Liangzhu town of Yuhang County (today's Yuhang District) in Zhejiang Province — and was named thereafter — where abundant jade ritual items were excavated and thus brought Liangzhu Culture to fame.

Jade wares of Liangzhu Culture were various in shapes: jade *Cong*, jade *Bi*, jade *Yue*, crown-shaped ware and trident-shaped ware, etc., and were mainly used as ritual objects to worship the Gods. Additionally, in this period, penetrated and sewn set ornament emerged, such as strung *Huang* (semi-annular jade pendant), tubes, beads and pendants. The common designs on jade wares varied from bird pattern, animal-faced pattern, god-human with animal-faced pattern, human-face pattern etc., which revealed man's worship of nature, totem and ancestors in prehistory. The shape and pattern of jade pursued a symmetric effect. And carving technique mainly focused on concave, combining with bass relief, round sculpture, semi-round sculpture and openwork etc.

● 玉三叉形器（良渚文化）
器物正面雕琢兽面纹，四周刻卷云纹，下部用浅浮雕琢出内有四枚獠牙的兽嘴。
Trident-shaped Jade Ware (Liangzhu Culture, 4000-5000 years ago)
It has animal-faced pattern carved in the front, curved cloud-shaped pattern all around, and an animal mouth with four fangs in bass relief on the lower part.

● 玉串饰（良渚文化）
Jade Ornament Set (Liangzhu Culture, 4000-5000 years ago)

夏商西周时期玉器

夏商西周时期，玉器已经彻底地从石器中分离出来，并开始成为显示地位、财富的象征物。帝王常常将玉器作为贵重物品赏赐给贵族、大臣，可见玉器已被赋予丰富的政治内涵。

Jades of the Xia Dynasty, the Shang Dynasty, and the Western Zhou Dynasty

In the Xia, Shang and Western Zhou dynasties, jade wares were completely segregated from stone wares and began to be regarded as the emblem of social status and wealth. Emperors often bestowed jade wares as precious awards

夏王朝是中国历史上第一个奴隶制国家，这一时期最重要的文化遗存是位于河南省偃师一带的二里头文化遗存。这里出土的玉器数量众多，集中出土于遗址墓葬中。出土的玉器以礼器、兵器为主，器形有玉戈、玉刀、玉钺、玉圭、玉璋、柄形器等。这些玉器表面大多光素无纹，只有个别器物上饰有弦纹、几何纹、兽面纹和人面纹等。器体大多长大宽薄，抛光莹润，象征着统治者的权力、地位和威严。

on aristocrats and officials. It is thus evident that jade wares had been endowed with a rich political connotation.

The Xia Dynasty (appox.2070 B.C.-1600 B.C.) is the first slavery state in Chinese history, whose cultural relics are located at Erlitou site in Yanshi area, Henan Province where considerable jade wares were excavated, especially from the ancient tombs. Ritual items and weapons constituted the main part of the unearthed artifacts, including jade *Ge* (sickle-shaped weapon), jade knife, jade *Yue* (axe-shaped weapon, turning into ritual object afterwards), jade *Gui* (belt-shaped ritual object with the bottom being flat and straight, and the upper part being triangle-shaped or straight), jade *Zhang* (belt-shaped ritual object with the bottom being flat and straight, and the upper part being knife-edged on one side like a half-*Gui*) and other handle-shaped jade wares. Their surface was plain without any motifs, only seldom of which had string pattern, geometric pattern, animal-faced pattern or human-face pattern as ornamental decoration. These ritual objects were usually long, large, broad and thin, well polished, symbolizing the power, status and majesty of the ruler.

- 玉圭（夏）
  Jade *Gui* (Xia Dynasty, appox.2070 B.C.-1600 B.C.)

- 玉璋（夏）
  夏代玉器中常见"出齿"装饰，即在玉器的轮廓上雕刻出凹凸的齿牙。
  Jade *Zhang* (Xia Dynasty, appox.2070 B.C.-1600 B.C.)
  The jade ware of Xia Dynasty was usually applied with protruding teeth decoration—carving out the jagging teeth on the rim of the jade ware.

商王朝尚玉，中国玉器在这一时期发展达到第二个高峰期。商文化一般以盘庚迁殷为标志分为早晚两期。

商早期玉器，以河南省郑州市二里岗出土的玉器为代表。玉器造型简单，主要是玉礼器和象征性的玉工具、玉兵器，纹饰较少。商早期玉器中最突出的是形体较大的玉戈，有的长近一米。商晚期玉器，以河南安阳殷墟出土的玉器为代表。殷墟出土的玉器多达2000余件，其中妇好墓就出土精美玉器755件，造型丰富，纹饰精美，人物、动物形象生动传神。

- 鸟首形玉戈（商）
Bird-headed Jade *Ge* (Shang Dynasty, 1600 B.C.-1046 B.C.)

- 玉戈（商）
Jade *Ge* (Shang Dynasty, 1600 B.C.-1046 B.C.)

The Shang Dynasty (1600 B.C.-1046 B.C.) worshiped jade to a great extent and also witnessed the second summit of jade ware development in China. The event of Pan Geng (the nineteenth king of Shang Dynasty) moving the capital of the Shang Dynasty to Yin (near the present-day Anyang City) around 1320 B.C. divided the Shang culture into early and late stages.

Jade wares in the early period of the Shang Dynasty, represented by those excavated at Erligang site in Zhengzhou City, Henan Province, have simple shapes with little decoration, mainly including jade ritual items, jade tools and weapons, among which jade *Ge* (sickle-shaped weapon) stands out with larger contour sometimes as long as one meter. Jade wares in the late Shang Dynasty are widely distributed in Anyang City of Henan Province where the Yin Dynasty Ruins (Yin Xu, the first recorded relics of a capital city in the Shang Dynasty) are located. More than two thousand pieces of jade were excavated there and the famous Fuhao Tomb alone contained 755 exquisite jade artifacts variously molded with attractive patterns and designs, as

背景知识

盘庚迁殷：商朝建立时，最早的都城在亳（今河南商丘）。后来因为商朝王位竞争，以及自然水灾等原因曾五次迁都，盘庚迁殷是其中的一次。商王盘庚即位时，商朝政治混乱，为解决政治危机他决定迁都至殷。殷，即殷墟，自盘庚迁都于此至商王朝灭亡，整个商朝后期都以此为都。

妇好墓：妇好是商代帝王武丁的妻子。妇好墓于1976年发掘，墓室不大，但保存完好，随葬品极为丰富，是研究商文化的重要遗存。

Background Knowledge

Pan Geng Moving the Capital of the Shang Dynasty to Yin: The original capital of the Shang Dynasty was located at Bo (modern Shangqiu City, Henan Province). Later, kings of the Shang Dynasty moved the capital altogether five times because of throne fighting and floods, among which Pan Geng was best known for moving the capital to Yin. When he came to the throne, the tumultuous political condition forced him to move to Yin and thus saved the Shang Dynasty from crisis. Yin, also known as Yin Dynasty Ruins, remained capital till the end of the Shang Dynasty.

Fuhao Tomb: Fuhao was the wife of the Shang king, Wuding (1259 B.C.-1201 B.C.). Her tomb discovered in 1976, has a burial chamber not very large but well preserved with abundant funeral objects as important cultural relics for the further research of the Shang Dynasty.

- 两面玉人（妇好墓出土）

玉人呈站立状，两面雕刻图像不同，一面为男子像，一面为女子像。

Two-faced Jade Figure (Excavated in Fuhao Tomb)

This human standing statue has two faces differently carved with male and female images.

- 玉凤（妇好墓出土）

玉凤作回首欲飞状，亭亭玉立，其造型基本符合后人对凤的描绘。

Jade Phoenix (Excavated in Fuhao Tomb)

The jade phoenix turns its head and is about to fly, posturing in slim and graceful style, possessing all the basic features of phoenix described by later generations.

周部落原是商王朝的属国。周武王灭商后，建都镐京（今陕西省西安市西部），成立了西周王朝。西周是中国奴隶制的鼎盛时期，大力推行以王权为中心的礼制。而玉器作为礼仪用具被广泛使用，出现了以"六器"和"六瑞"为代表的形制不同、用途各异的礼玉。

- 玉鹰（西周）
  Jade Eagle (Western Zhou Dynasty, 1046 B.C.-771 B.C.)

- 玉鹿（西周）
  Jade Deer (Western Zhou Dynasty, 1046 B.C.-771 B.C.)

well as vivid human and animal images.

The Zhou tribe was originally a dependency of the Shang Dynasty and after the King Wu of Zhou (Zhou Wuwang) overthrew the Shang Dynasty, he established the Western Zhou Dynasty (1046 B.C.-771 B.C.) and made Haojing (in the west of present-day Xi'an City, Shaanxi Province) its capital. The development of Chinese slavery regime culminated in the Western Zhou Dynasty when ritual norms (ritual norms refers to *Li*, a concept celebrated in Chinese Confucian philosophy to emphasize the significance of social norms and stratum) centering upon royal power was enthusiastically embraced. As a result, jade wares were widely used in this period as ritual items at ceremonies, among which ritual jade artifacts emerged in various shapes and with different usages, represented by the "six ritual objects" and "six auspicious items".

Excavated jade wearing accessories made in the Western Zhou Dynasty (1046 B.C.-771 B.C.) include animal-shaped— fish, deer, bird, etc. — jade pendants, a set of jade pendants, and strung jade ornaments. Decoration in this period usually employed pattern or design of simple yet vigorous lines, meanwhile the intaglio with two juxtaposed notched

出土的西周时期的大量佩玉，有以鱼、鹿、鸟等动物为形状的玉佩，还有由不同的饰件连缀而成的玉组佩、玉串饰。西周玉器纹饰线条简练刚劲，流行双勾阴刻，并创造了斜刻（一面坡）的新手法。lines came to fashion, and moreover a new carving technique was invented, that is, slope carving (or one-side slope).

- 玉凤鸟纹戈形器（西周）

斜刻（一面坡）是指用两条并列的阴刻线来表现纹样的轮廓，然后在阴刻线的一侧用斜刻法，将阴线的沟槽斜着磨宽一些，让纹样凸出，有立体感。

Jade *Ge* with Phoenix Design (Western Zhou Dynasty, 1046 B.C.-771 B.C.)

Slope carving (or one-side slope) means to apply two parallel concave lines to delineate the contour of the decorative pattern and use slope carving on one side of the concave lines to broaden the grooves by grinding aslant, so that the pattern could be highlighted and simultaneously allow a stereoscopic vision.

- 玉组佩（西周）

Jade Pendant Set (Western Zhou Dynasty, 1046 B.C.-771 B.C.)

六器、六瑞

六器：是周天子向天地神灵表示虔诚之意的玉器，是最高级别的玉礼器，包括苍璧、黄琮、青圭、赤璋、白琥、玄璜。

六瑞：是代表天命王权的信物，也是最高级别的玉礼器，包括镇圭、桓圭、信圭、躬圭、谷璧、蒲璧。

Six Ritual Objects, Six Auspicious Items

Six Ritual Objects: the highest-level jade ritual items specifically designed for kings of the Zhou Dynasty to worship heavenly deities and earthly spirits, including Green *Bi* (round and plate-shaped jade ritual object with a hole in the center), Yellow *Cong* (ritual object, round inside and square outside), Blue *Gui* (belt-shaped ritual object with the bottom being flat and straight, and the upper part being triangle-shaped or straight), Red *Zhang* (belt-shaped ritual object with the bottom being flat and straight, and the upper part being knife-edged on one side like a half-*Gui*), White *Hu* (tiger-shaped jade ware, usually used in ritual ceremonies), Black *Huang* (semi-annular jade pendant).

Six Auspicious Items: the highest-level jade ritual items regarded as pledges of the mandate of heaven and royal majesty, used by the six ranks of emperor, duke, marquis, count, viscount, baron, which respectively were *Zhen Gui*, *Huan Gui*, *Xin Gui*, *Gong Gui*, *Gu Bi* (millet pattern *Bi*), *Pu Bi* (cattail pattern *Bi*).

- 苍璧 Green *Bi*
- 黄琮 Yellow *Cong*
- 青圭 Blue *Gui*
- 赤璋 Red *Zhang*
- 白琥 White *Hu*
- 玄璜 Black *Huang*

春秋战国时期玉器

春秋战国时期，中国社会由奴隶制向封建制转化。随着周制"礼崩乐坏"，玉的礼仪功能逐渐减弱，佩玉之风盛行。加上各诸侯国制作的玉器风格竞美，综合运用各种技法制作玉器，玉器的发展进入第三个高峰期。这一时期，和田玉大量输入中原，玉的影响范围有所扩大。儒家把礼学与玉联系起来，认为玉的特性体现出儒家的仁、义、礼、智、信、乐、忠等传统观念。

Jades of the Spring and Autumn Period and the Warring States Period

The Spring and Autumn Period and the Warring States Period witnessed a remarkable transformation in Chinese history: the replacement of slavery by feudalism. As the social system established upon ritual norms collapsed, the ritual function of the jade ware deteriorated, while it widely prevailed as an ornament. Along with the competition of jade producing among different states, people created different styles and combined various techniques to refine their jade

- **孔子像【局部】马远（宋）**

  孔子（前551—前479），名丘，字仲尼，春秋末期鲁国人，中国古代伟大的思想家、教育家、儒家学派的创始人。他提出"君子比德于玉"的观点，认为玉具有十一德，成为中国人几千年以来爱玉、惜玉、佩玉风尚的理论依据。

  Portrait of Confucius (Partially, by Ma Yuan, Song Dynasty, 960-1279)

  Confucius (Kong Qiu, 551 B.C.-479 B.C.), styled Zhong Ni, was born in the Chinese State of Lu (in present-day Shandong Province) at the end of the Spring and Autumn Period and is known as a great thinker, educator, and founder of the Confucian school. He raised the concept to compare the virtue of a gentleman to jade which has altogether eleven virtues (benevolence as *Ren*, wisdom as *Zhi*, ritual as *Li*, righteousness as *Yi*, music as *Yue*, loyalty as *Zhong*, honesty as *Xin*, heaven as *Tian*, earth as *Di*, virtue as *De*, principles as *Dao*), providing the theory base for Chinese to love, cherish and wear jade over centuries.

春秋玉器在山西、陕西、山东、河南、江苏等地都有出土，其有代表性的器形有玉璜、玉觿、玉琥、玉圭、玉环、玉璧、玉玦、玉佩等。常见的纹饰题材有人物、龙、凤、虎等。有的扁平玉器上的两面纹饰不同，是春秋玉器纹饰最为突出的特点。

- 玉双龙首璜（春秋）
Jade *Huang* with Two-dragon-heads Design (Spring and Autumn Period, 770 B.C.-476 B.C.)

- 龙凤合体玉佩（春秋）
Jade Pendant with Dragon-phoenix Design (Spring and Autumn Period, 770 B.C.-476 B.C.)

战国玉器以湖北、河南、河北等地墓葬中出土的玉器最具代表性。其中湖北随县曾侯乙墓出土的玉器器形包括琮、璧、璜、环、瑗、玦、佩、

artifacts. Jade wares reached the third summit in development during this period of time. Meanwhile, Khotan jade was largely transported into the Central Plains, which made the culture of jade widely spread. The Confucian tended to associate ritual norms (*Li*) with jade, holding that the latter in a sense embodied some traditional ideas, in particular the Confucian virtues like benevolence (*Ren*), righteousness (*Yi*), ritual (*Li*), wisdom (*Zhi*), honesty (*Xin*), music (*Yue*), loyalty (*Zhong*), etc.

Jade wares of the Spring and Autumn Period (770 B.C.-476 B.C.) were excavated in many places like Shanxi, Shaanxi, Shandong, Henan, Jiangsu Provinces, represented by jade *Huang*, jade *Xi* (awl, a tool designed to untie the knots), jade *Hu*, jade *Gui*, jade ring, jade *Bi*, jade *Jue* and jade pendant, usually decorated with motifs of human figures, dragon, phoenix and tiger. The most conspicuous feature of jade wares in this period is that some flat jade artifacts have two faces in different patterns.

Jade wares of the Warring States Period (475 B.C.-221 B.C.) are best represented by those excavated from tombs in Hubei, Henan and Hebei Provinces, among which the tomb of marquis Yi of Zeng at Suixian County of Hubei Province

梳、带钩、觽、剑饰等。河南省辉县魏国墓出土有较完整的玉礼器，包括玉圭、玉璜、玉册、玉环等。此时的玉器纹饰流行几何纹。

has found jade artifacts molded in various shapes including *Cong*, *Bi*, *Huang*, ring, *Yuan* (ritual object with a larger hole than *Bi*), *Jue*, pendant, comb, belt hook, *Xi*, sword ornaments, etc. Several intact jade ritual items were excavated from the Tombs of Wei State at Huixian County in Henan Province, including jade *Gui*, jade *Huang*, jade book (*Ce*) and jade ring. Besides, geometric patterns prevailed in this period.

- 三龙璧形玉佩（战国）

玉佩出土于河北平山县古中山墓。墓中出土玉器3000余件，大多制作精致。图中的三龙璧形玉佩是一件珍品，在璧的外缘分布有三条相同的"S"形神龙，具有创新意义。

Bi-Shaped Jade Pendant with Three Dragons Coiling Around (the Warring States Period, 475 B.C.-221 B.C.)

It was excavated from the ancient Zhongshan tomb in Pingshan County of Hebei Province. More than 3000 pieces of jade were unearthed here and most of them were exquisitely made. The jade pendant illustrated is a rarity of innovation with three similar S-shaped divine dragons coiling around the outer rim of the *Bi*.

- 青玉带钩（战国）

Gray Jade Belt Hook (Warring States Period, 475 B.C.-221 B.C.)

- 卷龙形玉佩（战国）

玉佩出土于湖北随县（今随州）曾侯乙墓。墓中出土了大批精美的随葬品，其中玉器200余件。

Coiled Dragon-shaped Jade Pendant (Warring States Period, 475 B.C.-221 B.C.)

This jade pendant was excavated in the tomb of marquis Yi of Zeng at Sui County (present-day Suizhou) of Hubei Province, together with other exquisite funeral objects, including more than two hundred artifacts jade wares.

- 丝束纹玉环（战国）

Jade Ring with Pattern of Bound Silk (Warring States Period, 475 B.C.-221 B.C.)

秦汉玉器

秦汉时期，是中国封建社会的成长时期，也是玉器发展的全新时期。特别是汉代通西域以后，丝绸之路的建立，为新疆和田玉大量进入中原地区提供了方便。玉器制作具备了充足的优质原料，玉器的使用也更为广泛，兴起了佩玉、葬玉、用玉印章的风尚。汉代时，玉器的发展进入又一个高峰期。

- 勾连云纹青玉高足玉杯（秦）
Gray Jade Cup with Curved Cloud-Shaped Pattern (Qin Dynasty, 221 B.C.-206 B.C.)

Jades of the Qin Dynasty and the Han Dynasty

The Qin and Han dynasties witnessed the growth of Chinese feudal society as well as a brand-new development of jade wares. Especially, the Silk Road was established after the Han Dynasty reached the Western Regions (Xiyu, ancient name of nations and regions on the west of Yumen Pass and Yang Pass), which made it convenient to transport Khotan jade in great amount from Xinjiang to the Central Plains. Plenty materials of high quality facilitated jade making, impelling a wider use of jade wares. The custom of wearing jade, using jade as funeral objects and making seals started to prevail. As it were, during the Han Dynasty jade wares reached another summit in development.

Among the jade wares of the Qin Dynasty, *Heshi Bi* and the imperial jade seal are of the most fame. A batch of jade artifacts were excavated at the ancient city and palace of the Qin Dynasty in Shaanxi Province, including jade *Huang*, jade *Bi*, jade *Gui* and jade *Zhang*, of which a few were exquisitely made like the gray jade cup with curved cloud-shaped pattern.

Considerable jade wares were excavated from tombs of the Han Dynasty (206 B.C.-220 A.D.) in Mancheng

背景知识

和氏璧的故事

春秋时期,楚国人卞和在荆山(今湖北南漳县)发现了一块璞玉,便想将玉献给楚厉王,但玉工辨认后认为这是一块毫无价值的石头,厉王将卞和以欺君之罪斩去了左脚。楚武王即位后,卞和再次将玉献上,结果武王以相同罪名将其右脚斩去。等到楚文王即位后,卞和不敢再轻易进宫献玉,就怀抱璞玉在荆山脚下痛哭流涕。文王听说后,派人前去询问缘由,卞和说:"我不为失去双足伤心,伤心的是宝石被说成寻常石头,忠诚被说成是欺骗。"文王得知后立刻派人迎回璞玉,令人剖开,结果真是块稀世美玉,这块玉被称为"和氏璧"。秦始皇统一六国后,和氏璧由玉匠精心雕琢成价值连城的传国玉玺,上刻"受命于天,既寿永昌"八个篆字,成为帝王至高权力的象征。五代后唐废帝李从珂被契丹击败,持玉玺登楼自焚,玉玺自此下落不明。

传国玺

历朝在开国之初,都会用玉做一方印,由历任皇帝流传使用,是皇帝身份、权力的象征,这就是传国玉玺。

- "大清受命之宝"玉玺(清)
  Imperial Jade Seal with Words of "Treasure of the Great Qing Dynasty with Mandate of the Heaven" (Qing Dynasty, 1616-1911)

Background Knowledge

Story of *Heshi Bi*

The legendary story starts with a man named Bian He who found an uncut jade stone in the hills of Jing in Chu State (present-day Zhang County, Hubei Province) and presented it as a tribute to King Li of Chu. The jade artisans, however, identified it as valueless and the king therefore had his left foot cut off as punishment for deception. When King Wu, King Li's son came into power, Bian He presented the stone again, but likewise King Wu did not believe the man and had his right foot amputated as well. When King Wen of Chu succeeded the throne, Bian He dared not present the jade any more but shed bitter tears at the foot of hills with the jade in his bosom. Soon King Wen heard of it and sent a person to ask why. "I do not feel depressed for my loss of feet," said Bian He, "but for a precious gem mistaken as a common stone and loyalty misjudged as deception." Informed of the reason, King Wen immediately sent for Bian He and ordered for the stone to be cut by his sculptors. Surprisingly, the stone turned out to be one of the rarest jade ever to be discovered. It was made into a *Bi* (round and plate-shaped jade ritual object with a hole in the center) and named "*Heshi Bi*" — literally meaning "The *Bi* of He" — in honor of its discoverer Bian He. After the Emperor Ying Zheng (known as the Qin Shi Huang of the Qin Empire) conquered the other six warring states, the jade stone was carefully made into a priceless imperial seal by sculptors. Words "Having received the mandate from the heaven, may (the emperor) enjoys a long and prosperous life" were carved onto the seal which became the emblem of royal supremacy. The seal was lost in the Later Tang Dynasty (923-936) when the dethroned emperor Li Congke, defeated by Khitan, took it with him into the flames and committed suicide in the palace.

Imperial Seal

At the beginning of each dynasty, a seal would be made out of jade and later passed onto the following emperors, symbolizing the high status and supreme power of the emperor and is thus known as imperial seal.

• 仿制的传国玉玺印文
Imitation of the Imperial Jade Seal Print

秦代玉器中曾有著名的"和氏璧"和"传国玺"。在陕西境内的秦代古城和宫殿中出土有一批玉器，有玉璜、玉璧、玉圭、玉璋等，其中有少量精致之作，如勾连云纹高足玉杯。

汉代玉器出土较多，在河北省满城县（今满城区）、广东省广州市、北京、湖南省长沙市、河南省洛阳市、陕西省咸阳市、山东省莱芜市、安徽省亳州市、广西壮族自治区贵港市，以及山东省荣成市等地的汉墓都有发现玉器。汉代玉器数量庞大，种类也较为丰富。汉代玉器可分为礼玉、葬玉、陈设玉器、装饰玉四类。

汉代玉器纹饰以几何纹和动物纹为主，几何纹有谷纹、蒲纹、涡纹和云纹。动物纹有抽象纹饰和写实纹饰两种，抽象动物纹主要有龙纹、鸟纹、兽面纹、螭虎纹等，写实动物纹既有神话传说中的神仙怪兽，也有人物鸟兽等。汉代玉器雕刻手法多样，普遍使用透雕、高浮雕，并开创独具特色的"汉八刀"工艺。

District, Hebei Province; Guangzhou City, Guangdong Province; Beijing; Changsha City, Hunan Province; Luoyang City, Henan Province; Xianyang City, Shaanxi Province; Laiwu City, Shandong Province; Bozhou City, Anhui Province; Guigang City, Guangxi Zhuang Autonomous Region; and Rongcheng City, Shandong Province, etc. A large number of artifacts of rich variety exhibit the development of jade wares in this period which could be generally divided into four groups: ritual and funeral jade items, jade artifacts for display and ornament.

Geometric and animal patterns prevailed in the Han Dynasty (206 B.C.-220 A.D.). The geometric patterns included millet pattern, cattail pattern, swirl pattern and cloud pattern, and the animal patterns were divided into two categories: abstract designs like dragon pattern, bird pattern, animal-faced pattern, and dragon-tiger pattern (*Chi-Hu* pattern), and realistic designs like legendary deities, monsters, human figures, birds and animals. Besides, the diversified carving techniques in the Han Dynasty (206 B.C.-220 A.D.) also refined jade making, such as the widely-used openwork, high relief, and the ingenious carving method named *Hanbadao* (referring to the craftsmanship of shaping an object with several cuts).

汉代玉器分类
Classification of Jade Wares in the Han Dynasty

礼玉 传统礼玉中的璧和圭仍作为礼器，璜和琥只作为佩饰，琮和璋在汉代很少见，不作为礼玉使用。其中玉璧有素面璧、几何纹璧、多重饰纹璧、出廓式璧、铭文璧等几种。

Ritual Jade Wares: *Bi* and *Gui* were still used as ritual jade wares, whereas *Huang* and *Hu* were only made for wearing, also *Cong* and *Zhang* were rare in the Han Dynasty (206 B.C.-220 A.D.) and not used as ritual artifacts. And the *Bi* could be divided into five categories including plain *Bi*, geometric-pattern *Bi*, multi-layer-pattern *Bi*, *Chukuo* (a jade sculpting, literally means cross the outline, indicating an advanced style with bulging openwork dragon, phoenix or tiger design carved along the rim of the jade ware) *Bi*, and inscription *Bi*.

- 龙凤纹出廓玉璧（汉）
  Jade *Bi* with *Chukuo* Dragon-phoenix Design (Han Dynasty, 206 B.C.-220 A.D.)

- 犀牛纹玉璜（汉）
  Jade *Huang* with Rhinoceros Design (Han Dynasty, 206 B.C.-220 A.D.)

葬玉 在汉玉中占很大比例，是专门用来保存尸体的玉器，主要有玉衣、玉塞、玉蝉和玉握。

Funeral Jade Wares: Occupying a large proportion in Han jade wares, funeral jade items were specifically used to preserve corpses, like jade suit, jade *Sai* (also called *Jiuqiao Sai*, used to block up a corpse's nine orifices which are two eyes, two earholes, two nostrils, the mouth, genitalia and anus), jade cicada (funeral object placed inside the dead's mouth) and jade *Wo* (funeral object placed in the dead's hands).

• 金缕玉衣（汉）
Jade Suit Sewn with Gold Thread (Han Dynasty, 206 B.C.-220 A.D.)

陈设玉 有高足玉杯、玉耳杯、玉奁、卮、玉洗、玉枕、玉屏风、玉奔马、玉辟邪等。

Jade Wares for Display: There are high-foot jade cup, jade eared cup, jade *Lian* (cosmetics case), *Zhi* (cylindrical wine vessel), jade washer, jade pillow, jade screen, jade galloping-horse and jade *Bixie* (mythical animal, lion-like with wings), etc.

• 白玉仙人奔马（汉）
White Jade with Deity Riding Running Horse Design (Han Dynasty, 206 B.C.-220 A.D.)

• 玉豹（汉）
Jade Leopard (Han Dynasty, 206 B.C.-220 A.D.)

装饰玉 主要是佩玉，有由璧、璜、琥等与串珠组成的玉组佩，还有心形玉佩（也称"韘形佩"）、玉舞人佩、玉翁仲、玉笄、玉簪、玉带钩、玉剑饰，以及挂在腰带上的玉印和玉刚卯等。

Jade Wares for Ornament: It mainly includes jade pendants such as a set of *Bi*, *Huang*, and *Hu* strung together with beads, heart-shape pendant (also called She-shape pendant), jade pendant with dancing figure, jade *Wengzhong* (*Wengzhong*, famous warrior in the Qin Dynasty, the original prototype of the jade figure statue), jade *Ji* (hairpin), jade *Zan* (hairpin), jade belt hook, jade sword ornaments, jade seal and *Gangmao* (cuboid amulet with auspicious inscription carved on the four sides, can be strung up through the hole in the center) usually tied to the waist belt.

- 白玉双龙连体佩（汉）
  White Jade Pendant with Two-dragon Design (Han Dynasty, 206 B.C.-220 A.D.)

- 龙凤纹玉环（汉）
  Jade Ring with Dragon-phoenix Design (Han Dynasty, 206 B.C.-220 A.D.)

汉八刀

汉代玉器的一种雕琢技法，也是评价汉代玉器雕琢风格的术语，它的原意是一种注重造型轮廓神态的写意风格。汉代玉器的造型粗犷，仅用几刀就雕刻完成了，故称"汉八刀"。汉代葬玉中的玉握（玉猪）、玉蝉等器物常用此技法。

Hanbadao

It is one of the jade carving techniques in the Han Dynasty (206 B.C.-220 A.D.) and also a term to evaluate the carving style of the jade wares in this period, originally referring to a freehand style which emphasizes the contour and the expression of the objects. Bold and straightforward in shape, the jade wares of the Han Dynasty were only finished within several strokes, and thus "*Hanbadao*" acquired its name. As it were, funeral jade items such as jade *Wo* and jade cicada, were usually carved by this technique.

- 玉蝉（汉）
  Jade Cicada (Han Dynasty, 206 B.C.-220 A.D.)

魏晋南北朝及隋唐五代玉器

魏晋南北朝是中国玉器发展史上的低潮期，玉器做工和质地都明显逊于汉代。这一时期政局动荡，战乱不息，当时的士大夫流行服食玉屑，对玉器的发展产生了负面影响，玉器的种类和数量都减少了许多，礼玉和葬玉基本绝迹。装饰玉器逐渐成为主流，品种有玉蝉、玉杯、玉盏、玉辟邪、玉带饰、玉佩等。

到了隋唐五代时期，玉器受到与西域交流融合的影响，风格呈现出融合性。如与西域风格饰物宝珠

Jades of the Wei, Jin, Southern and Northern Dynasties, the Sui Dynasty, the Tang Dynasty, and the Five Dynasties

The development of jade wares slowed down during the Wei, Jin, Southern and Northern dynasties when both the craftsmanship and quality of artifacts were inferior to those of Han Dynasty (206 B.C.-220 A.D.). The tumultuous political situation, unceasing wars, together with the prevailing fashion of eating jade scraps among the literati and officialdom, brought about a negative effect on jade development. As a result, jade wares reduced in variety and quantity, ritual and funeral jade items almost disappeared, while ornamental jade wares predominated such as jade

● 黄玉瑞兽（魏晋南北朝）
Yellow Jade with Auspicious Animal Design (Wei, Jin, Southern and Northern Dynasties, 220-581)

● 人狮纹玉带板（魏晋南北朝）
Jade Belt Tablet with Human-lion Design (Wei, Jin, Southern and Northern Dynasties, 220-581)

搭配组合的玉龙，形似云南、贵州等地的孔雀的玉凤，饰有胡人纹饰的玉带等。这一时期玉器的文化内涵开始转向现实的世俗生活，龙凤开始从神灵转为帝后的象征。

隋代因存在时间较短，出土的玉器也较少，目前只有李静训墓出土的几件玉器，比较著名的有金口白玉盏、玉兔、玉镯、玉钗等。

唐代出土的玉器虽也不多，但比较能够代表这一时期玉器的总体特点，具有代表性的玉器造型有玉带板、玉飞天等。

- **击鼓乐伎纹玉带板（唐）**
  唐代玉带板纹饰以人物、花鸟、动物为主，其中人物形象多为胡人。
  Jade Belt Tablet with Musician Beating Drum Design (Tang Dynasty, 618-907)
  Tang Dynasty mainly employed figures, mainly *Hu* people (ancient name of the non-Han minority nationality, called by the Han people), flowers, birds and animals as the ornamental design for jade belt tablet.

cicada, jade cup, jade *Zhan* (small cup), jade *Bixie* (mythical animal, lion-like with wings), jade belt ornaments, jade pendant, etc.

When it came to the Sui, Tang and the Five dynasties, jade wares were greatly influenced by the cultural communication and interaction with Western Regions, and thus achieved an integrative style, like the jade dragon combined with precious pearl of Western Regions' style, the jade phoenix resembling the peacock of Yunnan and Guizhou Provinces and the jade belt decorated with patterns of Western Regions' style, etc. Additionally, the cultural connotation of jade wares in this period turned to embrace the present worldly life, hence dragon and phoenix were no longer divine spirits but symbols of king and queen.

Due to the short duration of the Sui Dynasty (581-618), the unearthed jade wares are in small quantity. So far only several jade artifacts are excavated from the tomb of Li Jingxun, including the famous white jade *Zhan* (small cup) with gold mouth, jade hare, jade bracelet, and jade *Chai* (hairpin).

Likewise, the unearthed jade wares of the Tang Dynasty are in small quantity, but still being representative of the overall characteristics of the jade wares in this

五代十国时期出土的玉器更少，种类也只有玉哀册、玉步摇、玉带，以及少量的玉坠、玉佩饰等，纹饰则流行龙凤鸟图案。

period, such as jade belt tablet and jade *Feitian*.

Fewer jade wares of the Five dynasties and the Ten states (907-979) are excavated, among which there are only jade book of eulogy (*Aice*), jade dangling hairpin (*Buyao*), jade belt as well as a few jade pendants and accessories, usually with prevalent designs of dragon, phoenix and bird.

- 金口白玉盏（隋）
White Jade *Zhan* with Gold Mouth (Sui Dynasty, 581-618)

- 玉飞天（唐）
玉飞天为女人形象，上身裸露、赤脚，肩上披一条长飘带。
Jade *Feitian* (Tang Dynasty, 618-907)
The *Feitian* normally is a female image with naked upper-body and bare feet, draping a long ribbon over her shoulders.

宋辽金元玉器

宋代（960—1279）以前，玉器一直是统治阶级的专有用品。而

- 玉罗汉（宋）
Jade Buddhist Arhat
(Song Dynasty, 960-1279)

- 玉云鹤纹饰件（宋）
Jade Ornament with Cranes-and-clouds Design (Song Dynasty, 960-1279)

Jades of the Song Dynasty, the Liao Dynasty, the Jin Dynasty and the Yuan Dynasty

Before the Song Dynasty (960-1279) jade wares were exclusively used by the governing class. When it came to the Song Dynasty, however, owing to the development of economy and commercial trade, the craftsmanship of jade-making gradually went out of the royal court and spread among the folk, and meanwhile the variety of jade wares also progressed toward secularization. Despite the frequent warfare and quick dynasty changes in this period, the jade carving techniques still moved forward. Besides, affected by the prevailing royal court vintage style, sculptors of the Song Dynasty began to replicate ancient jade wares, which developed into a certain scale and also produced some exquisite jade artifacts of high quality. Yet only a small number of them are excavated, among which more than ten items unearthed from the stone coffin of Changgouyu tomb in Fangshan District, Beijing are generally regarded as the representatives of jade wares in the Northern Song Dynasty (960-1127), whereas items excavated from the tomb of Shi Min(ying)zu

春水玉和秋山玉

春水玉描绘的是辽、金、元代帝王贵族在春季围猎的场景,以海东青(一种猛禽)攫天鹅为题材,通常用镂雕技法制作而成。画面中海东青体态小巧,矫捷勇猛,天鹅则惊慌失措,仓皇躲藏,周围点缀花草,风格写实,具有强烈的民族特色。

秋山玉描绘的是辽、金、元代帝王贵族在秋季围猎的场景,以虎扑鹿为题材,通常采用镂雕工艺来雕琢山、林、虎、鹿等自然画面,或单面雕,或双面雕,充满了山林野趣,极富草原游牧民族的特色。

Spring-water Jade (*Chunshui* Jade) and Autumn-hills Jade (*Qiushan* Jade)

Spring-water jade is carved in openwork with the theme of *Haidongqing* (one kind of falcon) seizing the swan, delineating a spring hunting scene participated by the emperor and noblemen of the Liao Dynasty (907-1125), the Jin Dynasty (1115-1234) and the Yuan Dynasty (1206-1368). The *Haidongqing* in the picture is small, nimble and vigorous, while the swan, panic-stricken, tries to hide itself in a rush. Embellished around are flowers and grass, which together with the falcon and swan, are realistic in style and highlight distinctive ethnic features.

Autumn-hills jade depicts a similar autumn hunting scene, which adopts tiger springing at deer as its theme, usually applies openwork to carve hills, woods, tiger and deer, on one side or both sides of the jade, vividly characterizing the attractive wilderness and full of nomadic features.

- 春水玉饰(元)
Spring-water Jade (*Chunshui* Jade) Ornament (Yuan Dynasty, 1206-1368)

- 秋山玉炉顶(金)
Autumn-hills Jade (*Qiushan* Jade) Burner Crest (Jin Dynasty, 1115-1234)

到了宋代，随着经济与商业贸易的发展，玉器工艺才逐渐从宫廷走到民间，玉器种类也朝着世俗化、生活化的方向发展。宋辽金元时期，虽然朝代更迭，战争频繁，使玉器生产受到了一定影响，但制玉技术仍然在不断地进步。另外，受到宫廷复古风的影响，宋代开始出现仿古玉的制作，并形成了一定的规模，其中不乏质优形美的精品玉器。宋代玉器出土的不多，在北京房山（今房山区）长沟峪石椁墓中出土的十几件玉器，被认为是北宋玉器的代表。浙江省衢州市史绳祖墓出土的玉器，则是南宋玉器的代表。宋代玉器以人物、花卉、飞禽、童子等纹样为主，采用镂雕技法雕刻。

辽（907—1125）、金（1115—1234）、元（1206—1368）时期的玉器一方面受到宋代玉器的影响，另一方面又具有北方游牧民族的特色，在中国古代玉器中别具一格。最具代表性的题材是"春水玉"和"秋山玉"。

(1192-1274) at Quzhou City, Zhejiang Province are characteristic of jade wares in the Southern Song Dynasty (1127-1279). Song sculptors usually employed openwork for jade-carving in patterns of human figures, flowers, birds and children (*Tongzi*).

Influenced by those of the Song Dynasty on the one hand and tinted with northern nomadic color on the other, jade wares in the Liao Dynasty (907-1125), the Jin Dynasty (1115-1234) and the Yuan Dynasty (1206-1368) were ingenious with a distinctive style. The most representative themes were Spring-Water jade (*Chunshui* jade) and Autumn-Hills jade (*Qiushan* jade).

Jades of the Ming Dynasty and the Qing Dynasty

During the Ming and Qing dynasties, as the economy and commercial trade became more prosperous, jade-making also enlarged in scale, gradually commoditized and specialized. Besides, the quantity and variety of jade wares used by the royal family — usually for daily use, display, ornament or appreciation — in this period surpassed any other previous dynasties, whereas among the common people it also became

明清玉器

明清时期，随着经济与商品贸易的更加繁荣，玉器制作的规模不断扩大，并向商品化方向发展，玉器生产日益专业化。这一时期皇家用玉的数量和品种超越了以往任何一个时期，用玉范围包括日用、陈设、佩饰、文玩等方面。而民间的赏玉、玩玉也十分盛行，文人、富商都热衷于收藏、品玩玉器。

明代玉器种类包括宫廷礼仪用玉、器皿、摆件、佩饰和各式实用玉器。另外还有仿古的宫廷礼仪用

a trend to appreciate jade artifacts, and scholars and rich merchants were quite enthusiastic about jade collection and appreciation.

Jade wares of the Ming Dynasty (1368-1644) include royal ritual objects, vessels, displays, ornaments and various kinds of practical jade artifacts, as well as imitations of ancient ritual objects such as jade *Bi* and jade *Gui*, mainly decorated by studs pattern and *Chi*-dragon pattern. Bold and vigorous in style, jade wares in this period widely used openwork, yet left the parts of minor importance such as the interior and the bottom roughly managed, which made jade wares of the Ming Dynasty known as "coarse Great Ming" (*Caodaming*). However, there was no lack of exquisite jade artifacts as well,

• 兽面蕉叶纹耳杯（明）

明代玉器的制作承袭了元代北方工匠琢玉的特点，制作出的作品多粗犷、大气、宽厚，常常忽略细部的处理，工艺略显粗糙。

Animal-faced Jade Eared Cup with Banana-leaf Pattern (Ming Dynasty, 1368-1644)

Jade wares of the Ming Dynasty inherited the technical features of northern sculptors in the Yuan Dynasty (1206-1368) and therefore were bold, vigorous and generous, often ignoring the management of details. As a result, the craftsmanship appeared slightly coarse.

玉，如玉璧和玉圭等，纹饰多为乳钉纹和螭纹。此时的玉器造型粗犷浑厚，广泛使用镂雕技术，但对内腔、底部等次要部位处理不细致，有"糙大明"的说法。明代玉器中也有精致之作，以苏州琢玉高手陆子冈的作品最负盛名。

清代是中国玉器最为发达的时期，出现了众多玉器作坊，其中一些实力强大的专业作坊由宫廷直接控制，经常接受宫廷钦定的琢玉任务。清代玉器品种齐全，而且几乎每一类玉器都有惊世之作。翡翠的加入也为清代玉器增光添彩，无论是在用料、工艺、艺术表现方面，清代玉器都达到了历史最高的水平。

among which best known are the works of Lu Zigang, a highly-skilled sculptor in Suzhou City.

The Qing Dynasty (1616-1911) witnessed the most prosperous period in jade development. Abundant jade workshops were established, among which the powerful and professional ones were under the direct control of the royal court and often received imperial order. Jade wares were rich in variety and there were amazing pieces almost in each kind. The addition of jadeite further brought luster to the jade wares of the Qing Dynasty which, as it were, reached the highest level of development in terms of materials, craftsmanship and artistic expression in Chinese history.

陆子冈

陆子冈，也作"子刚"，是明嘉靖后期至万历初期的苏州琢玉高手，善雕各种器物，纹饰、器形往往是仿古再加以改造，风格秀雅精致。

陆子冈所制玉器在当时很有名，是当时官僚、皇室追捧的收藏品。但明代后期，民间玉器行出现仿造名家之风，故有"子刚"款识的很多，款识很杂乱，但绝大多数为后人所制，其真品很难见到。北京故宫博物院和台北故宫博物院收藏有刻有陆子冈款识的仿古玉杯、青玉执壶、合卺杯、臂搁、印盒、琴式盒和佩等。

Lu Zigang

Lu Zigang was a highly-skilled jade sculptor in Suzhou City during the late Jiajing Period and early Wanli Period of the Ming Dynasty. Expert in carving all kinds of wares, he imitated and reformed patterns and shapes of ancient artifacts, and finally achieved a delicate and refined individual style.

Jade wares carved by Lu Zigang were very famous at that time, pursued and admired by officials and royal family as valuable collections. However, in the late Ming Dynasty, folk sculptors began to imitate the works of jade masters and therefore many artifacts were marked with "Zigang" among which the few were genuinely made by the very person. The Palace Museums in Beijing and Taibei possess jade wares marked with "Lu Zigang", such as the imitating antique jade cup, gray jade handled ewer, paired cup (*Hejin* cup, nuptial cup with bound double-cup shape), arm-rest, seal case, *Guqin*-shaped case (*Guqin* refers to a seven-strung ancient Chinese harp) and pendant, etc.

- "子刚"款白玉龙首簪（明）
  White Jade *Zan* (Hairpin) with Dragon-headed Design and Zigang Mark (Ming Dynasty, 1368-1644)

- "子刚"款玉合卺杯（明）
  Paired Jade Cup (*Hejin* Cup) with Zigang Mark (Ming Dynasty, 1368-1644)

清代玉器分类
Types of Jade Wares in the Qing Dynasty

器皿类 主要有杯、碗、盘、壶、盒、盏、盂、盅，以及仿古的鼎、簋、尊、觥、觚、爵等。

Vessels: It mainly includes cup, bowl, plate, ewer, case, *Zhan* (small cup), *Yu* (broad-mouthed liquid container), *Zhong* (small cup without handle), as well as imitation wares such as *Ding* (cauldron), *Gui* (round and eared food container), *Zun* (goblet), *Gong* (an ancient wine vessel made of horn), *Gu* (an ancient wine vessel with broad mouth and slender waist) and *Jue* (three-legged drinking vessel with loop handle), etc.

- 青玉描金题诗碗（清）
  Gray Jade Bowl with Gold Poem Inscription (Qing Dynasty, 1616-1911).

- 翡翠瓜形壶（清）
  Jadeite Pumpkin-shaped Ewer (Qing Dynasty, 1616-1911).

陈设类 主要有玉山子、插挂屏和一些形制较大的儒、佛、道造像，瑞兽等。许多大型玉山子的纹饰以山水为主，点缀以人物、建筑、车船等，其中著名的有"大禹治水图"玉山子、"秋山行旅图"玉山子、"赤壁泛舟图"玉山子等，都是以名画家的画稿为蓝本进行设计制作的。

Jade wares for display: It mainly includes jade *Shanzi* (jade display with landscape design), table screen, hanging panel, some large-sized Confucian, Buddhist and Taoist statues and auspicious animals. Many of the large jade *Shanzi* have landscape as the primary pattern and are occasionally embellished with figure, house, carriage and boat, among which particularly notable are jade *Shanzi* with Yu the Great Regulating Water design, jade *Shanzi* with Autumn Travel in Mountains design and jade *Shanzi* with Boating at Red Cliff (*Chibi*) design. All of them were ingeniously carved based on the sketches of famous painters.

- 玉观音（清）

Jade Avalokitesvara (*Guanyin*) Statue (Qing Dynasty, 1616-1911)

- "大禹治水图"玉山子（清）

"大禹治水图"玉山重10000余斤，是用一块巨大的青白玉山料雕成。玉山作重山叠岭、古木参天、流水飞瀑之状，在险峻的悬崖绝壁上，大禹带领着成群结队的民工在开山导石，在山巅浮云处，一位神仙带着几个鬼怪，仿佛在开山爆破，使这件描写现实的作品具有了浪漫主义的色彩。

Jade *Shanzi* with Yu the Great Tames the Flood Design (the Qing Dynasty, 1616-1911)

More than 5,000kg in weight, this jade *Shanzi* was carved out of a greenish-white jade (Qingbai jade) material and decorated with folds of mountains, towering ancient trees, flowing water and plunging waterfall, among which Yu the Great leads a crowd of people quarrying above the steep cliffs. Together with them is a celestial being amidst clouds at the mountain peak who guides several sprites as if to open the mountains by demolition, which further grants a romantic color to this realistic picture.

装饰类 主要有朝珠、翎管、项圈、手镯、戒指、带钩、环、佩、钗、簪，以及各种镶嵌玉。

Jade Ornament: It mainly includes court beads (*Chaozhu*), feather tube (*Lingguan*), necklace, bracelet, ring, belt hook, hoop, pendant, *Chai* (hairpin), *Zan* (hairpin), and other inlaid jade ornaments.

- 玉簪（清）
  Jade *Zan* (hairpin) (Qing Dynasty, 1616-1911)

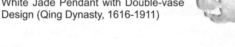

- 白玉双连瓶佩（清）
  White Jade Pendant with Double-vase Design (Qing Dynasty, 1616-1911)

文房类 包括笔筒、笔架、墨床等文房用品，以及印章、棋子等。

Jade Study Tools: It includes brush pot, brush rack, ink rest, as well as seal and chessman, etc.

- 青玉水仙花笔筒（清）
  Gray Jade Brush Pot with Narcissus Design (Qing Dynasty, 1616-1911)

- 白玉雕螭双联章（清）
  White Jade Double-seal with Design of Carved *Chi*-dragon (Qing Dynasty, 1616-1911)

礼器类 多为仿古的琮、璧、玉册等。另外，玉如意成为清代较为常见的礼器。

Ritual Jade Objects: It mainly includes imitations of ancient *Cong* (ritual object, round inside and square outside), *Bi* (round and plate-shaped jade ritual object with a hole in the center), jade book, etc. Jade *Ruyi* (a curved decorative object symbolizing good fortune) also became one of the most commonly seen ritual items in the Qing Dynasty.

- 青玉仿古勾云纹璧（清）
Gray Jade Imitation of Ancient *Bi* with Curved Cloud-shaped Pattern (Qing Dynasty, 1616-1911)

- 雕花玉如意（清）
Jade *Ruyi* with Carved Floral Design (Qing Dynasty, 1616-1911)

近现代玉器

Jade Wares of Modern Times

清末时期，清王朝的统治风雨飘摇，玉器加工却因为受到外国人的青睐而方兴未艾。但这一时期的玉器作坊由于属私人开设，财力有限，多以制作小件器物为主，很少有制作大件的材料。玉器制作风格以仿古、修改古玉为主，其形制、纹样基本沿袭明清风格，由于审美

At the end of the Qing Dynasty, the political turmoil drove the royal sovereignty into crisis, counter to which jade carving nevertheless prospered and enjoyed high favor from the foreigners. Private as most jade workshops were in this period, limited financial resources obliged them to give priority to small artifacts and few large-sized jade wares were produced for lack

观念日趋世俗化而显得过于庸俗、没有创新。

20世纪30年代，连年的军阀混战，加上抗日战争的爆发，刚刚有点复苏势头的玉器行业再次跌入谷底。玉器制作和加工停滞不前，只有民间玉雕有一点有限的成就。

中华人民共和国成立后，玉器行业恢复和发展得很快。北京和上海成为国内最重要的玉器制作中心，出现了一批玉雕工艺大师。此外，扬州、苏州、天津、广州等地

of materials. Besides, sculptors mainly replicated and reformed ancient jade; As a result, current jade wares basically inherited the formal and pattern design of the Ming and Qing dynasties and had little innovation because of the secularization of aesthetic taste.

In the 1930s, the recovering jade business again collapsed due to years of fighting among warlords and then the outbreak of the War of Resistance Against Japan (1937-1945). Except for the limited achievement in folk jade workshops, the whole business of jade

- **翡翠观音像（现代）**

  观音菩萨呈坐姿，脸部表情庄重，是北派玉雕作品。

  Jadeite Avalokitesvara (*Guanyin*)(Modern Times)

  The Avalokitesvara (*Guanyin*) is sitting upright with solemn facial expression. This piece of artifact belongs to the Northern School.

也是玉器制作的知名城市。当时，全国的玉器制作以北京和上海为中心分为南北两派。北派以北京、天津、河北等地为代表，其特点是人物造型类玉器都很注重表现人物的

production came to a standstill.

Jade business revived and developed rapidly after the People's Republic of China was founded (in 1949). Beijing and Shanghai were noted as jade production centers of that time where masters of jade sculpture emerged in large numbers. Besides, Yangzhou City, Suzhou City, Tianjin City and Guangzhou City were also well-known for jade-making. In this period nation-wide jade production generally fell into two groups: Beijing-centered Northern School and Shanghai-centered Southern School. The former was represented by jade wares of Beijing, Tianjin and Hebei Province, whose distinguished features could be identified from the vividly highlighted individual personalities of figure statues. The elegance of the maid, the vivacity of children, the solemnity of Bodhisattva, the ease and grace of deities, all were fully expressed. And jade wares modeling on flowers, birds, fish and worms were also vivid and delicate, and those with floral design were often realistically carved. Jade vessels, including jade *Xun* (a kind of incense burner to fume away ghosts and evil spirits), jade vase and jade censer, mainly inherited the formal style of the imperial court of the Qing

● 白玉宝塔炉（现代）
宝塔炉制作精巧，玲珑剔透，是南派玉雕作品。

White Jade Tower-shaped Censer (Modern Times)
Ingeniously and exquisitely carved, the tower-shaped censer is a production of the Southern School.

性格,仕女的秀美、童子的活泼、菩萨的肃穆、仙人的飘逸皆表现得淋漓尽致;仿花鸟鱼虫类玉器雕琢逼真细腻,花卉多写实;玉器器皿多沿袭清代宫廷造型,以熏、瓶、炉为主,薄胎和嵌宝压丝技术得到发展。南派以上海、扬州、苏州、广州为代表,其特点是玉器上的纹饰以写实为主,加工工艺上追求细腻、平滑,玉器器皿既有借鉴青铜器造型的,也有采用传统风格创造的,总体上给人以匀称、流畅、秀丽、高雅的美感。

而从20世纪50年代至20世纪80年代,中国玉器行业采用艺术品工厂体制进行管理,进一步推动了玉器行业的发展,提高了玉器制作的水平。这一时期中国主要的玉器厂培养了一批艺人,他们中大多数人成为当代著名的玉器制作大师,由他们设计制作或领衔制作的作品,质量很高,具有很大的投资收藏价值。近年来,由于和田籽料、翡翠藏量日趋稀少,导致玉石原料价格上涨,并带动整个新玉器市场价格的飙升,现代玉器大师的艺术作品更是成为市场中的抢手货。

Dynasty (1616-1911) and also benefited from the development of thin-body and gold-and-silver inlaying techniques. The Southern School represented by Shanghai, Yangzhou City, Suzhou City and Guangzhou City usually applied realistic patterns with delicate and smooth craftsmanship. Some of its jade vessels borrowed ideas from the bronze ware, while others adopted traditional style, however, manifested a beauty of balance, fluency, exquisiteness and elegance on the whole.

From the 1950s to the 1980s, the artwork factory management system was adopted by Chinese jade business, which not only promoted jade development but brought the jade craftsmanship onto a higher level. A number of sculptors emerged from major jade factories during this period of time and most of them were well known as jade carving masters, who designed and produced several jade artifacts of superb quality and high value for investment or collection. In recent years, the shortage of Khotan jade material and jadeite leads to the rise in the price of the jade material. As a result, the overall price of the new jade market soars and artifacts produced by modern jade masters are in great demand.

- **密玉攀登珠穆朗玛峰（现代）**
  Mi Jade with Design of Climbing the Mount Qomolangma (Modern Times)

  密玉，又名"河南玉"，因产于河南省密县而得名，属于中档玉料，常加工为摆件和首饰，颜色浓艳的用于制作戒指的戒面。密玉的质地、颜色、硬度、光泽变化较小，常以颜色、块度为主要分级标准。颜色好、块大的为上品，以绿中透翠的最为珍贵；颜色次、块小的为下品。

  Mi jade (also Henan jade), was named after Mi County, Henan Province, the place of its origin. Average in quality, *Mi* jade stones are usually carved into display or ornament, while those rich-colored ones are often used to make ring gem. Since there is little variation in texture, color, hardness and luster among *Mi* jade stones, sculptors apply color and lumpiness as grading standard. Thus accordingly top-grade items are those brilliant in color and massive in volume, whereas the ugly-colored and small ones belong to the low-grade.

> 玉的象征

从远古时代至今,中国人琢玉、用玉、爱玉的风气代代延续。玉在中国人的心中远不仅仅是"美石",更是被赋予了丰富内涵的象征物。在中国人眼里,玉是天赐尤物,藏于高山深处,历经风霜雨雪的磨砺,承天地之灵气,集日月之光辉。从最初对玉的图腾崇拜到以德比玉,再到寓意吉祥,玉一直是中国人心中至高无上的象征。

> The Symbol of Jade

Chinese generations have been continuing to carve, use and cherish jade from the remote ancient times up to the present. Jade is no longer simply a beautiful stone in the eyes of Chinese but a symbol endowed with rich connotation. Chinese tend to regard jade as a grace from heaven, a rarity hidden deep in mountains, tempered by harsh environment, bathed in the glory of the sun and the moon, and nourished by

- **玉鹰(凌家滩文化)**
  玉鹰呈振翅欲飞状,翅膀末端均做成猪头形,胸部雕刻一个圆圈,内有八角星纹,可能与祭祀太阳有关。

  Jade Eagle (Lingjiatan Culture, 5300-5600 years ago)
  The jade eagle flaps its wings and is about to fly, with pig-headed design at the end of the wings and a circle of octagonal star pattern carved in the chest, which may be used in the sun-god sacrificial ceremony.

玉是沟通神灵的媒介

中国远古祖先称玉为"美石",并且对"美石"进行挑选,加工成具有装饰功能的佩饰,玉还被认为是通灵之物,并被当作原始图腾而受到人们的崇拜。除此之

• **良渚文化玉钺上的神徽纹样**

神徽纹样造型奇特,神人冠羽冲天,双目圆瞪,龇牙咧口,双臂左右暴撑,兽面也同样圆瞪双目,威猛狰狞。神徽纹样表现的很可能是一种祖先崇拜。

Pattern of Divine Figure on Jade *Yue*
(Liangzhu Culture, 4000-5000 years ago)

Strange in design, it depicts a fierce divine figure with towering crest, round-opened eyes, teeth out and arched arms stretching outward with full strength, and an animal-faced pattern which is likewise ferocious with wide-dilated eyes. This design probably manifests ancestor worship among ancient Chinese.

heavenly and earthly spirits. Throughout the millennia the symbol of jade continuously enjoys a supreme position in Chinese mind, originally adored as a totem, then associated with human virtue, and later elevated as emblem of happiness and fortune.

Jade as a Medium for Communicating with Spirits

The ancient Chinese ancestors called jade beautiful stone and selected the best of it to be carved into ornamental pendant. Jade was also considered to be psychic (i.e. capable of communicating with ghosts or spirits) at that time and therefore was greatly adored by the ancients as primitive totem. Besides, the jade artifact bearing particular connotations would serve as emblem of clan in ancient times.

Primitive jade wares often had bird, animal, and deity as formal designs or had divine figure, animal-faced pattern carved onto the surface of the artifacts, which marked the possible beginning of primitive worship of jade totem. However, since the production and usage of jade at that time was limited, only the chief and the priest were well qualified to wear and use jade wares. Besides, jade often appeared on solemn occasions

外，人们还用带有某种含义的玉石制品作为氏族的标志。

原始玉器多被雕成鸟、兽、神人等造型，或在器表雕琢出神人、兽面等图案。这大概就是原始人类玉图腾崇拜的标志。而当时玉的开发与使用极少，只有族群里的族长、祭师才有资格佩戴和使用玉器。玉经常在庄重的场合出现并扮演重要的角色。

玉是地位与财富的象征

随着古代社会从氏族公社时期进入奴隶社会时期，私有制产生，玉就为少数氏族领袖所占有。自夏代以后，玉成为国家政治的重要内容，是尊贵身份的象征物，在重大的祭祀活动中作为礼器使用。到了商周时期，随着奴隶制国家的发展成熟，因政治需求，玉的地位越来越高。著名典籍《周礼》中记述了关于当时玉器的分类及使用规定。

在中国古代，玉一直集中在统治阶级和王公贵族组成的上层社会，各朝代的帝王都占有大量的玉器，并将之视为最贵重的财富。时至今日，玉虽然不再具有封建权势

and whereupon played a role of great significance.

Jade as Symbol of Social Status and Wealth

As the clan commune system was replaced by the slavery system in ancient China, private property came into being and henceforward jade was mainly possessed by a few clan chiefs. In the Xia Dynasty (approx. 2070 B.C.-1600 B.C.), jade began to play an important part in political field, became a symbol of noble status and constantly served as ritual object in significant sacrifices to heaven and earth, ghost and spirit. When it came to the Shang and Zhou dynasties, as the slavery state developed and matured, jade responded to the political needs and got higher social status. The well-known classic, *The Rites of Zhou Dynasty (Zhouli)*, recorded the classification and regulations on usage of jade wares at that time.

In ancient China, jade was only possessed by the upper society including the ruling class and the noblemen. Emperors of each dynasty owned abundant jade wares and regarded them as great treasure. Nowadays jade is no longer the symbol of feudal power but

的象征意义,但却因其稀有、不可再生的特性,依然具有极高的经济价值和收藏价值。中国有句俗话说"黄金有价玉无价",可见玉象征着无穷的财富。

still retains a high economic value and is worthwhile for collection because of its rarity and non-renewability. As an old Chinese saying goes, you can set a price for gold, but never to jade. It is thus evident that jade as a symbol bears infinite wealth.

- 玉权杖(良渚文化)

Jade Truncheon (Liangzhu Culture, 4000-5000 years ago)

- 夏禹王像 马麟(宋)

玉圭在古代用来表明身份等级,是权力的象征。画中夏禹王手执玉圭,表现出古代帝王至高无上的气势。

Portrait of Yu the Great of the Xia Dynasty (by Malin, Song Dynasty, 960-1279).

Jade *Gui* (a belt-shaped ritual object with the bottom being flat and straight, and the upper part being triangle-shaped or straight) is a symbol of power and was used in the ancient China to represent one's social status. As illustrated, with a jade *Gui* in hands Yu the Great radiates a supreme power of ancient Chinese emperor.

玉是美好品德的象征

春秋战国时期，和田玉开始大量输入中原，其坚硬、洁白、温润等特点被儒家用来比作仁、义、礼、智、信等传统美德。孔子提出"君子比德于玉"的观点，并将玉的品质与儒家倡导的道德伦理规范进行比照，作了详细的阐述。他认为玉所体现的德性共有十一种，即仁、智、义、礼、乐、忠、信、

● 彩绘木俑（战国）

古代贵族礼服上常于腰带或裙上系挂玉佩，用作装饰。木俑腰带上所系为玉组佩，自上而下为珠、管、瑗（环）、璜，穿缀而成。

Painted Wooden Figurine (Warring States Period, 475 B.C.-221 B.C.).

The formal dress of ancient noblemen often has a jade pendant as ornament tied to the waistband or the skirt. Such is the case with the wooden figurine in the picture who has a set of pendants attached to the belt, which is made up of bead, tube, loop, and *Huang* from top to bottom.

Jade as Symbol of Virtue

During the Spring and Autumn Period and the Warring States Period (770 B.C.-221 B.C.), Khotan jade was transported into the Central Plains in great quantity. It was solid, white and gentle in texture, therefore was used by the Confucian to embody traditional virtues such as benevolence (*Ren*), righteousness (*Yi*), ritual (*Li*), wisdom (*Zhi*), honesty (*Xin*). Confucius, a great ancient philosopher in China, once compared the virtues of a gentleman to that of jade. Through a detailed comparison between the quality of jade and the morality advocated by the Confucianism, he concluded that jade represented altogether eleven virtues, including benevolence (*Ren*), wisdom (*Zhi*), righteousness (*Yi*), ritual (*Li*), music (*Yue*), loyalty (*Zhong*), honesty (*Xin*), heaven (*Tian*), earth (*Di*), virtue (*De*) and principles (*Dao*), which almost covered all the moral ethics proposed by the Confucian School.

Xu Shen, a scholar in the Eastern Han Dynasty (25-220), defined jade in *Shuowen Jiezi* (a book explaining and analyzing Chinese characters) as follows, "Jade is the fairest of stones. It is endowed with the five virtues. Benevolence (*Ren*) is typified by

背景知识

何谓仁、智、义、礼、乐、忠、信、天、地、德、道?

夫昔者君子比德于玉焉,温润而泽,仁也;缜密以栗,知(智)也;廉而不刿,义也;垂之如队,礼也;叩之其声清越以长,其终诎然,乐也;瑕不掩瑜,瑜不掩瑕,忠也;孚尹旁达,信也;气如白虹,天也;精神见于山川,地也;圭璋特达,德也;天下莫不贵者,道也。

——孔子《礼记·聘义》

Background Knowledge

What does the eleven-virtue mean?

The wise have likened jade to virtue. For them, its gentleness and brilliancy represent benevolence (*Ren*); its perfect compactness and extreme hardness typify the sureness of wisdom (*Zhi*); its angles, which do not hurt, although they seem sharp, signify righteousness (*Yi*); it looks like a pendant when droops and hangs, implying ritual (*Li*); the pure and prolonged sound, which it gives forth when stricken and could stop abruptly, alludes to music (*Yue*); its flaws never obscure its splendor and vice versa, which calls to loyalty (*Zhong*); glittering, translucent and glorious in color, it bears the quality of honesty (*Xin*); its iridescent brightness represents heaven (*Tian*); its admirable substance, born of mountain and of water, represents the earth (*Di*); made into jade tablet, it is used at the royal court as a token of trust, indicating the virtue (*De*); and the price that the entire world attaches to it represents the principles (*Dao*).

By Confucius *Liji-Pinyi*

天、地、德、道,其原则和规范几乎涵盖了儒家道德规范。

东汉文学家许慎在《说文解字》中概括玉有五德,即"玉,石之美者,有五德:润泽以温,仁之方也;䚡理自外,可以知中,义之方也;其声舒扬,专以远闻,智之方也;不挠而折,勇之方也;锐廉而不忮,洁之方也。"他把玉石在

its luster, bright yet warm; righteousness (*Yi*) by its translucency, revealing the color and markings within; wisdom (*Zhi*) by the purity and penetrating quality of its note when the stone is struck; courage (*Yong*), in that it may be broken but cannot be bent; honesty (*Jie*), in that it has sharp angles which yet injures none." Thus he associated the excellent qualities of jade in texture, luster, hardness, vein, tone

质地、光泽、硬度、纹理、音色等方面的优良品质，与封建社会的道德规范和个人的品德修养紧密地联系在一起。这成为中国人爱玉、佩玉的精神寄托，是几千年玉文化的重要内容。

color, etc., with the moral integrity of an individual. Those who love and wear jade find spiritual sustenance from Xu Shen's viewpoint which indeed constitutes an important content in Chinese jade culture throughout the millennia.

"玉"字的演化

"玉"本是象形字，呈几片玉穿在一起的形状，但与"王"字容易混淆，于是人们在"王"字的基础上加了一点，以示与"王"字的区别。

The Evolution of the Chinese Character *Yu* (玉, Jade)

Originally the character *Yu* was pictographic in the shape of several pieces of jade strung together, and therefore was similar to another Chinese character *Wang* (王 , king). In order to distinguish one from another, people added a dot to the character *Wang* (王 , king) and thus got the present character *Yu* (玉 , jade).

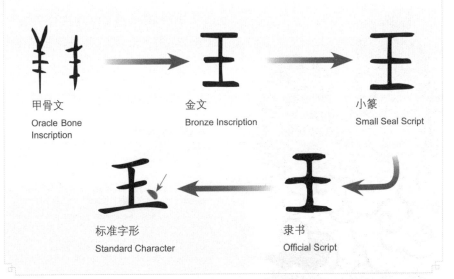

玉是吉祥如意的象征

从唐宋开始，玉器趋于世俗化，渐渐不再为上层社会所专有，而是走向民间。随着民间使用的普及，佩饰、把玩器（能握在手里触摸和欣赏的玉器）成为玉器的主流。其造型、纹饰多表现人们追求美好生活、祈求平安幸福的主题。玉器的题材和种类也不断扩大，多运用人物、动物、植物，以及一些民间谚语、吉语和神话故事，通过借喻、比拟、双关、象征、谐音等表现手法，构成"一句吉语，一副图案"的表现形式，成为吉祥、美好的象征。

Jade as Symbol of Happiness and Fortune

Since the Tang and Song dynasties, jade wares had been gradually secularized. No longer possessed exclusively by the upper class, they became part of the folk life, among which jade accessories and small jade artifacts (held in hand for touch and appreciation) enjoyed great popularity. The shape and pattern represented people's pursuit of peace and happiness. Besides, jade wares were also enriched with theme and variety. Figure, animal, plant, as well as folk proverb, auspicious word and fairy tale were vividly exhibited by means of metaphor, analogy, pun, symbolism and assonance, etc. The combination of auspicious words and corresponding image rendered the jade article a symbol of happiness and fortune.

● **碧玉鱼龙形花插（清）**
花插为鱼龙形，取"鲤鱼跳龙门"（鲤鱼跳过龙门就能变成龙）的典故，寓意金榜题名、步步高升。

Green Jade Flower Receptacle with Fish-dragon Design (Qing Dynasty, 1616-1911)
This flower receptacle with fish-dragon design, which alludes to the story of Carps Jumping over the Dragon Gate (for those carps which jumped over the dragon gate can turn into the real dragon), symbolizes eminent official career with continuous promotion.

玉与山水文化

人类由山水孕育而生，自始就与山水相依存，崇拜自然、追求天人合一，是人的天性使然。从魏晋时期的纵情于山水，到明清时期的寓情于园，人们寄情山水，而玉便是表现山水文化的一种载体。明清时期，随着玉器雕琢工艺的发展，在玉牌、玉山子、玉插屏等玉器中，以山水风光、文人隐逸生活为题材的，屡见不鲜，成为这一时期玉器装饰的主题之一。

Jade and Landscape Culture

Nurtured by the landscape, human race from the very beginning depended on nature for existence. They worshiped it and strived to live in harmony with it and such is human nature. From the Wei, Western and Eastern Jin dynasties when people indulged themselves in mountains and waters, to the Ming and Qing dynasties when they found emotional sustenance in gardens as well as landscapes, while jade happened to become a carrier for the expression of landscape culture. As the craftsmanship of jade carving got improved in the Ming and Qing dynasties, jade wares with themes of landscape or hermit life were quite common and even became the main trend in that period of time, including jade tablet, jade *Shanzi* (jade display with landscape design) and jade table screen, etc.

- 会昌九老图玉山（清）

Jade Mountain with the Design of Nine Elders Getting Together in Huichang (Qing Dynasty, 1616-1911)

- 人物山水纹玉香筒（清）

Jade Incense Tube (Aroma Diffuser) with Design of Figures and Landscape (Qing Dynasty, 1616-1911)

玉与酒文化

　　我国酒文化是一种特殊的文化形式，有着丰富的内涵，玉制酒具便是酒文化的载体之一。精美别致的酒具，造型独特，集实用性与观赏性于一身。玉杯是酒具中最主要的器物，始见于西汉，当时有筒状、角形等不同造型。唐代时，饮酒之风盛行，玉杯的运用更为广泛。明清时，玉杯外表多有大量花纹装饰，更加具有观赏价值。

Jade and Wine Culture

Distinctive as it is, Chinese wine culture is endowed with rich connotation and partially represented by the jade wine vessels, which, exquisite and unique in shape, is designed both for practical use and aesthetic enjoyment. Jade cup outshines above all as the most important vessel. It first appeared in the Western Han Dynasty and was molded into tube or horn shape at that time. When it came to the Tang Dynasty, wine drinking prevailed as a custom and led to a wider use of jade cup. The Ming and Qing dynasties decorated jade cup with abundant floral patterns on the surface and thus attained higher visual value.

- 玉角形杯（汉）
Jade Horn-shaped Cup (Han Dynasty, 206 B.C.-220 A.D.)

- 青白玉花形杯（明）
Greenish-white Jade Cup with Floral Design (Ming Dynasty, 1368-1644)

玉与印文化

　　我国印文化历史久远，秦代以前玉与金、银、铜是制印的主要材料，所制之印称"方寸玺"，人人都可使用。秦代规定只有皇帝的印章独称玺，专以玉制成。玉材质珍贵，篆刻艺术风格独特，不同的玉印造型还体现了拥有者的身份和地位。西汉时，高祖刘邦授予诸侯使用玉印的权利。从玉印的传世数量来看，汉代居多，战国次之，秦代玉印非常少。魏晋时期，由于印章艺术风格开始转变，玉印制作减少。隋唐以后，除皇家以玉琢制玺印外，其他多为玩赏之物。

Jade and Seal Culture

Chinese seal culture can be traced back before the Qin Dynasty, when seal was primarily made of jade, gold, silver and bronze, which was named as square imperial seal (*Fangcun Xi*) and could be used by anyone. In the Qin Dynasty, however, it was prescribed that only the seal used by emperor could be called imperial seal (*Xi*) exclusively, which was specifically made by rare jade material and was molded by unique carving technique into various shapes to represent the social status of its holder. Emperor Gaozu (also Liu Bang) of Western Han Dynasty empowered the feudal princes to use jade seal. Judging from the amount of jade seals handed down from ancient times, Han Dynasty harvested the maximum, followed by the Warring States Period, and then by the Qin Dynasty when few jade seals was produced. Because of a transformation in style, jade seals decreased in number during the Wei, Western and Eastern Jin dynasties. Ever since the Sui and Tang dynasties, jade had been carved mainly for enjoyment only with exception of the royal family who still made to use jade seals.

- 玉印（西汉）

Jade Seal (Western Han Dynasty, 206 B.C.-25 A.D.)

- "八徵耄念之宝"玉玺（清）

Imperial Jade Seal Designed to Celebrate the Eightieth Birthday of Emperor Qianlong (Qing Dynasty, 1616-1911)

中国玉器中常见吉祥图案

观音：佛教菩萨之一，形象端庄慈祥，有普度众生、大慈大悲、救苦救难、吉祥如意等寓意。

弥勒佛：形象为笑口大张、双耳垂肩、袒胸露腹的老者形象，有笑口常开、大肚能容、知足常乐等寓意。

寿星：形象为高脑门、白发白须的老翁，寓意长寿。

和合二仙：中国民间传说中的两位神仙，一持荷花，一捧圆盒，盒内盛满珠宝，并飞出一串蝙蝠，寓意财富无穷尽。荷、盒与"和合"同音，多比喻家庭和睦、夫妻团圆。

关公：三国人物，商人通常将他看作监察诚信、辟邪消灾的守护神。

钟馗：中国民间传说中能驱除鬼邪的神。相貌奇异，但正气凛然，寓意避邪消灾。

仙鹤：形象高贵、洒脱、洁雅，是延年益寿的象征。

白菜：汉语中"菜"与"财"谐音，寓意招财、聚财。

葡萄：呈硕果累累之态，象征丰收，寓意事业有成。

船：象征一帆风顺。

• 玛瑙和合福寿（现代）
Agate *He-He Fu Shou* (Happiness and Longevity) Statue (Modern Times)

Common Auspicious Designs of Chinese Jade Ware

Avalokitesvara (*Guanyin*): Avalokitesvara is a Bodhisattva in Buddhism. Decent and kind in appearance, she is also infinitely merciful and compassionate to deliver all the living creatures from misery and bring them happiness and good fortune.

Maitreya: It is a grinning elder with mouth wide open, long ears hanging over the shoulders, chest and belly laid bare. He is a symbol of optimism, magnanimity, and self-contentment.

God of Longevity (*Shouxing*): It is an elder with high forehead, white hair and white beard, signifying longevity.

Two Immortals of *He-He* (Harmony and Unity): They are two legendary immortals in Chinese folklore. One holds a lotus flower in hand, while the other carries a round box filled with precious stones and from which a train of bats fly out, indicating endless wealth. The lotus (*he*, 荷) and the box (*he*, 盒) share the same sound with *He-He* (和合), denoting the harmony and unity of a family.

Guan Gong: He is a famous historic figure in the Three Kingdoms (220-280) and is usually regarded by the merchants as a tutelary god of loyalty and righteousness.

Zhong Kui: He is a legendary god in Chinese folklore who though queer in appearance yet reveals awe-inspiring rectitude and supposedly protects man from evil spirits.

Divine Crane: It is elegant, pure and free, also a symbol of longevity.

Chinese Cabbage: It symbolizes the accumulation of wealth given the assonance of the character *Cai* (菜, cabbage) and *Cai* (财, wealth).

Grapes: With the fruitful image, grapes suggest good harvest and great achievement in one's career.

Boat: It signifies a nice trip.

• 翡翠"百事如意"
Jadeite Chinese Cabbage Signifying All Going Well as One Wishes (*Baishi Ruyi*)

关于玉的常用成语

玉成其事：指成全某件好事，多用于男婚女嫁。
玉骨冰肌：形容女子苗条的身材和光润洁白的肌肤。
玉洁冰清：像玉那样纯洁，像冰那样清净。形容人心地纯洁，品行端正。
玉树临风：形容男子像树一样风度潇洒。
金玉良言：比喻可贵而有价值的劝告。
金童玉女：原是道家侍奉仙人的童男童女，后泛指天真无邪的男孩女孩。
金枝玉叶：原形容花木枝叶美好，后比喻出身高贵或娇嫩柔弱的人。
金玉良缘：比喻美好的婚姻。
琼浆玉液：指美酒或甘美的饮料。
琼楼玉宇：原指仙界的亭台楼阁，后形容华美的楼宇、宫殿。
锦衣玉食：华美的服饰和美味的饭食，形容奢侈豪华的生活。
雕栏玉砌：形容富丽堂皇的建筑物。

Common Idioms about Jade

Yucheng qishi: It means to assist another in accomplishing a task, especially that of marriage.
Yugu bingji: It depicts the slender figure and fine skin of a female.
Yujie bingqing: It illustrates an honest and noble person who is as pure as jade and as clean as ice.
Yushu linfeng: It likens a handsome and elegant male to a jade tree.
Jinyu liangyan: It signifies invaluable advice.
Jintong yunv: It originally means boy and girl attendants of fairies in Taoism and now generally refers to the innocent and pure boys and girls.
Jinzhi yuye: It originally describes the beauty of flowers and trees, and now refers to those born of noble family or delicate and frail in condition.
Jinyu liangyuan: It is used to symbolize happy marriage.
Qiongjiang yuye: It indicates excellent wine or other sweet drinks.
Qionglou yuyu: It originally refers to the heavenly palace and is now used to depict magnificent building or palace.
Jinyi yushi: It means gaudy dress and delicious food, suggesting an extravagant life.
Diaolan yuqi: It is used to describe magnificent and splendid buildings.

玉之美
The Beauty of Jade

　　玉在中国人心中是美的化身。历代制玉者把自然美、精神美注入玉石之中，塑造了众多造型多样、纹饰精美的玉器。玉质美、造型美、纹饰美构成了多彩多姿的美玉世界。

Jade is the incarnation of beauty in the eyes of Chinese. Jade sculptors of each generation added natural and spiritual beauty to the jade stone and produced enormous jade wares of rich variety and exquisite design. The fine material, together with the excellent molding and ingenious design, makes a splendid and colorful jade world.

> 名玉之美

中国是世界上玉的分布地域最广、产量最大、开采历史最悠久的国家。中国名玉荟萃，和田玉、独山玉、岫玉、绿松石、酒泉玉、昆仑玉、翡翠等，举不胜举。

> The Beauty of Famous Jade

As the greatest jade producer in the world, China has the longest history of jade mining and widest deposits area. Famous jade such as Khotan jade, *Dushan* jade, *Xiu* jade, turquoise, Jiuquan jade, Kunlun jade, jadeite, etc., assemble in China and are too numerous to enumerate.

玉的概念

在1997年5月1日正式实施的《珠宝玉石名称国家标准》中，玉石被定义为："由自然界产出的，具有美观、耐久、稀少性和工艺价值的矿物集合体，少数为非晶质体。"其中"矿物集合体"是指岩石。

中国玉器行业认定的"玉"的种类很多，包括和田玉、翡翠、岫玉、绿松石、孔雀石、玛瑙、水晶、芙蓉石、珊瑚、琥珀等。

Definition of Jade

On May 1st 1997, *the National Standard for Naming of Jewelry and Gem* took effect, according to which jade is defined as natural mineral aggregate, with a few amorphous ones. It is beautiful, durable, rare, and endowed with technical and artistic value.

Various types of jade are certified by Chinese jade business, such as Khotan jade, jadeite, *Xiu* jade, turquoise, malachite, agate, crystal, rose quartz, coral, amber, etc.

和田玉

和田玉，又名"新疆玉""昆仑玉"，主要产于新疆维吾尔自治区塔里木盆地之南的昆仑山北麓，成矿带延续长达1100多千米。和田玉是中国传统玉器制作的主要原料，与河南的独山玉、辽宁的岫玉、湖北的绿松石一起被称为"中国四大名玉"，其中尤以和田玉最受推崇。新石器时代良渚文化遗址中曾发现用和田玉制成的玉琮。宫廷玉器多用和田玉制成，所以和田玉又有"帝王玉"之称。

Khotan Jade

Khotan jade (also named Xinjiang jade, Kunlun jade) is mainly mined in the mineral belt, more than 1,100 km in length, at the northern foot of the Kunlun Mountains, to the south of the Tarim Basin in the Xinjiang Uygur Autonomous Region of northwestern China. It is the primary traditional material in China and known as "The Four Famous Jades of China" together with *Dushan* jade of Henan Province, *Xiu* jade of Liaoning Province, and turquoise of Hubei Province, among which Khotan jade enjoys greatest reputation. In Liangzhu Cultural (4000-5000 years ago) site of the

- **昆仑山**
在中国古代，昆仑山被认为是万山之祖，它高大雄伟且盛产美玉，受到人们的崇拜。
Kunlun Mountains
In ancient China, Kunlun Mountains were regarded as the forefather of all other mountains. Magnificent in appearance and abundant with jade, the mountains were greatly venerated by Chinese.

和田玉属造岩矿物闪石族中的透闪石——阳起石，主要化学成分是硅酸钙、镁、铁。和田玉在生成时因混有其他物质而呈现不同的颜色和质地。和田玉有韧性，耐磨，不易打出断口，断面呈参差状，在锯割和雕刻时不会断裂。和田玉具有冬天不冰、夏天不热的特点。和田玉一般呈微透明状，颜色纯正，温润有泽，抛光后具油脂光泽。和田玉的摩氏硬度为6～6.5度，密度为$2.90～3.02 g/cm^3$，折射率为1.61～1.63。

Neolithic Age, people once found Khotan jade *Cong* (ritual object, round inside and square outside). Besides, the court jade wares were mostly made of Khotan jade which, as a result, is also called imperial jade.

Khotan jade is a tremolite-actinolite of the rock-forming amphibole group that is a silicate of calcium, magnesium and iron. Mixed with other substances during its generation, Khotan jade usually appears in different color and texture. It is tough, endurable, and will not easily fracture, whereas the cross-section looks uneven and therefore will not break when cut or sculpted. Besides, Khotan jade possesses the feature of slow heat-conducting. It is a slightly transparent jade of pure color, gentle gloss and mild

- 《天工开物》中的白玉河

  白玉河发源于昆仑山北坡，盛产羊脂白玉、白玉、青玉和墨玉。籽玉就是原生玉矿经冲刷滚入白玉河，经过河水的冲洗磨炼而成的。

  White Jade River Illustrated in *Exploitation of the Works of Nature(Tiangong Kaiwu)*(by Song Yingxing, first published in 1637, Ming Dynasty)

  The White Jade River originates from the northern hillside of the Kunlun Mountains and abounds with mutton-tallow white jade, white jade, gray jade and black jade. Seed jade (*Ziyu*) is the primary jade ore brought down to the White Jade River after years of erosion and flushing.

和田玉有三种天然产出形状，代表了三种不同质地的玉料，即籽玉、山流水玉和山料玉。籽玉是指埋藏较浅的和田玉矿石，因自然风化、冰川、泥石流、河水冲刷等自然原因被冲到河床下游，呈鹅卵石状，表面光滑，大小不一，是最佳的玉料；山流水玉是和田玉矿石经自然风化后，被泥石流、雨水冲刷搬至河流中上游形成的，表面呈较为光滑的次棱角片块状，质地介于籽玉与山料玉之间，属优良料种；山料玉是指直接从山上开采的和田玉矿石，往往呈不规则棱角块状，大小不一，质地一般不如籽玉好。

texture and has an oil-like luster after polishing, with the hardness of Mohs 6-6.5, density of 2.90-3.02 g/cm^3 and refraction index of 1.61-1.63.

Khotan jade has three output appearances, representing three materials with different textures, including seed jade (*Ziyu*), *Shanliushui* (literally means water flowing from the mountain) jade, and *Shanliao* (mountain material) jade. Seed jade is not deeply buried and then carried down to the lower reaches of the river by efflorescence, glaciers, debris flow and river water, with cobble-shaped, smooth surface and different in size, which is considered the best jade material. *Shanliushui* jade is formed from the weathered Khotan jade ore carried by debris flow and rainwater to the upper and middle reaches of the river, with relatively smooth exterior and an edged schistose-massive shape, and a texture between seed jade (*Ziyu*) and *Shanliao*

- 籽玉
  Seed Jade (*Ziyu*)

- 山流水玉
  Shanliushui Jade

- 山料玉
  Shanliao Jade

颜色也是决定和田玉品质的因素之一。和田玉按颜色大致可分为羊脂玉、白玉、青玉、黄玉、墨玉、碧玉等。羊脂玉颜色洁白，宛如羊脂，质地纯净，是和田玉中最名贵的品种；白玉质地纯正，是和田玉中比较名贵的品种，色越白越好；青玉以青色为主，质地细腻，

(mountain material) jade, considered the fine jade material. *Shanliao* (mountain material) jade refers to the Khotan jade ore which is mined directly from the mountain, presenting an irregular edged massive shape and various sizes, but is generally lower in quality than seed jade (*Ziyu*).

Color is another crucial element to identify the quality of Khotan jade, in terms of which Khotan jade can be divided into mutton-tallow jade, white jade, gray jade, yellow jade, black jade and green jade, etc. Mutton-tallow jade is the first-class Khotan jade, white as coagulated mutton grease and pure in texture. White jade is also pure in texture and often the whiter, the better,

- 碧玉卧牛摆件（明）
Green Jade Display with Design of Crouching Ox (Ming Dynasty, 1368-1644)

- 墨玉九龙三洗盆（清）
Black Jade Basin with Nine-dragon Design (Qing Dynasty, 1616-1911)

- 和田黄玉葫芦鼻烟壶（清）
Khotan Yellow Jade Snuff Bottle with Gourd Design (Qing Dynasty, 1616-1911)

颜色匀净，比白玉稍重，质地较白玉差；黄玉的黄色调是由于玉料长期受到氧化铁渗透而形成的，其品质与羊脂玉相近；墨玉颜色由墨色到淡黑色，全黑色的墨玉可用于制作器皿，散点状的墨色为脏色和杂质，黑白对比强烈的墨玉可做俏色作品；碧玉颜色有绿、深绿、暗绿色，以质地细腻、颜色纯正、具油脂光泽的墨绿色碧玉为上品。

considered as the precious kind of Kohtan jade. The gray jade is mainly in green-tone, with fine texture and evenly-distributed color, a bit heavier and worse in texture than the white jade. Yellow jade is close to the mutton-tallow jade in quality, and its yellow-tone is a result of the long-term infiltration of ferric oxide. The color of black jade varies from ink-black to light-black. The pure black jade can be used to make vessels. The scattered-dotted ink-black is dirty color and inclusions. And the black jade with strong black-white contrast can be carved into *Qiao*-color artifacts (*Qiao*-color is a technical term of jade-making, indicating

- 和田青玉兽面纹双耳瓶（清）
Khotan Gray Jade Vase with Double Ears and Animal-faced Pattern (Qing Dynasty, 1616-1911)

- 和田羊脂白玉"忠义千秋"摆件（现代）
Khotan Mutton-tallow White Jade Display with the Theme of Perpetual Loyalty (Modern Times)

• 和田白玉仕女（现代）
Khotan White Jade Maid Figure (Modern Times)

the color of a jade material is tactfully managed to integrate with the whole theme). The color of green jade also ranges from green, deep green to dark green, among which the top grade ones are those blackish green jade, pure in color, delicate in texture, with the gloss of mutton-tallow.

Dushan Jade

Dushan jade (also *Du* jade and Nanyang jade) originated from the Dushan Mountain in Nanyang City, Henan Province, which was known as Yushan Mountain (Jade Mountain) during the Western Han Dynasty (206 B.C.-25 A.D.). *Dushan* jade has been exploited and used ever since the Neolithic Age (4500-8500 years ago) and some *Dushan* jade wares such as shovel and chisel were excavated at the Longshan Cultural (2500 B.C.-1900 B.C.) site. Some of the *Dushan* jade articles also unearthed

独山玉

独山玉，又称"独玉""南阳玉"，产于河南省南阳市北郊的独山，西汉时独山被称为"玉山"。独山玉的开采和使用历史悠久，新石器时代龙山文化遗址中曾出土了独山玉制成的玉铲、玉凿等。河南殷墟妇好墓出土的众多玉器中，也有很多独山玉制品。独山当地玉器的生产至今仍在延续。

独山玉是斜长石类玉石，成分较为复杂，以斜长石和黝帘石为

• 渎山大玉海（元）

渎山大玉海是用一块黑质白章（黑色玉料上有白色的花纹）的河南独山玉雕琢而成，形制巨大，现藏于北京北海公园团城玉瓮亭内。

Dushan Jade *Hai* (Large Urn) (Yuan Dynasty, 1206-1368)

It is carved out of a white-patterned black Dushan jade, with a large size and is now preserved at the Jade Urn Pavilion of Tuancheng (circular city) in Beihai Park, Beijing.

主，其他成分还包括次闪石、透闪石——阳起石、透辉石，以及铬云母、角闪石、绿帘石、阳起石、葡萄石等。独山玉以细粒状晶体为主，质地细腻，坚硬致密，近似和田玉，但不如和田玉质地纯净。独山玉一般呈微透明至半透明状，有玻璃光泽，抛光面有油脂光泽。独山玉属多色玉石，单色的较为少见，一般有两种或两种以上颜色相互浸染交错。独山玉的摩氏硬度为6~8度，密度为2.73~3.18g/cm³，折射率为1.56~1.70。

from the Fuhao Tomb in the Yin Ruins of the Shang Dynasty (1600 B.C.-1046 B.C.). As it is, *Dushan* jade wares are continuously produced in its place of origin up to the present.

Dushan jade belongs to the plagioclase family and is mainly comprised of plagioclase and zoisite, as well as other substances including secondary amphibole, tremolite-actinolite, diopside, fuchsite, hornblende, epidote, actinolite, and prehnite, etc. *Dushan* jade is primarily composed by crystal particle, possessing a smooth texture, hard and compact, with a similar feature to the Khotan jade, yet less pure in quality. It appears slightly transparent or semi-translucent, with vitreous luster and mutton-tallow gloss on its polished exterior. *Dushan* jade is a polychrome jade stone, generally has two or more than two colors dipping with each other, and is rare in the single color. Its hardness is Mohs 6-8; the density is 2.73-3.18 g/cm³; and the refraction index is 1.56-1.70.

Color is a crucial element in identifying the quality of *Dushan* jade. According to its color, *Dushan* jade is divided into green *Dushan* jade, white *Dushan* jade, red *Dushan* jade, blue *Dushan* jade, yellow *Dushan* jade,

颜色是决定独山玉品质优劣的重要因素。独山玉按颜色不同可划分为绿独山玉、白独山玉、红独山玉、青独山玉、黄独山玉、褐独山玉、墨独山玉和杂色独山玉。浅色的玉质较好，深色的玉质较差。绿独山玉以绿、翠绿色为主，半透明，质地细腻，有玻璃光泽，近似翡翠；白独山玉以乳白色或灰白色为主，略透明，质地细腻，有油脂光泽；红独山玉又称"芙蓉玉"，呈粉红色至红色，透明性好，质地细腻，有玻璃光泽；青独山玉呈青绿色，不透明，是独山玉中较为常

brown *Dushan* jade, black *Dushan* jade and variegated *Dushan* jade. Generally speaking, the light-colored *Dushan* jade is higher in quality than the dark-colored one. Green *Dushan* jade resembles jadeite in quality, mainly green or bright green, semi-translucent, delicate in texture and with vitreous luster. White *Dushan* jade is mainly milk-white or off-white, slightly transparent, delicate in texture and with luster of grease. Red *Dushan* jade, also called lotus jade (*Furongyu*), varies from pink to red in color and has good transparency, with delicate texture and vitreous luster. Blue *Dushan* jade is one of the common types, opaque and with bluish green tone. Brown *Dushan* jade appears semi-translucent and in all shades of color, including light purple, purple or bright brown. Black *Dushan* jade is mainly black or in dark green, opaque and with rough and big crystal particle, belonging to the second-class *Dushan* jade. Variegated *Dushan* jade has two or more colors mixed together and is the commonest kind of *Dushan* jade.

- 独山玉"荷塘鹭鸶"摆件（现代）
  Dushan Jade Display with Design of Egrets in Lotus Pond (Modern Times)

见的品种；褐独山玉呈淡紫、紫色和亮棕色，颜色表现深浅不一，半透明。墨独山玉呈黑色或墨绿色，不透明，颗粒较粗大，是独山玉中质地不好的品种；杂色独山玉是指两种或两种以上的颜色混杂的独山玉，是独山玉中最常见的品种。

岫玉

岫玉又称"岫岩玉"，因主要产于辽宁省鞍山市岫岩县而得名。

• 独山玉"苏武牧羊"摆件（现代）
Dushan Jade Display with Design of Su Wu Herding Sheep (Modern Times)

Xiu Jade

Xiu jade (or *Xiuyan* jade) was named after its place of origin, Xiuyan County in Anshan City of Liaoning Province. During the Neolithic Age and Qin and Han dynasties, *Xiu* jade was the main material of jade wares. Some artifacts excavated at primitive sites such as that of Hongshan Culture (5000-6000 years ago) and Liangzhu Culture (4000-5000 years ago) are believed to be made of *Xiu* jade. The Fuhao Tomb of the Shang Dynasty (1600 B.C.-1046 B.C.) in the Yin Dynasty Ruins also has *Xiu* jade wares unearthed. The Jade Suit Sewn with Gold Thread discovered in the Tomb of Zhongshan Prince Jingwang (Liu Sheng) in Mancheng of Hebei Province, is also a product of *Xiu* jade. *Xiu* jade enjoys the highest output at present and occupies approximately 60 percent of China's total production volume of jade material every year.

• 岫玉松鹤瓶（现代）
Xiu Jade Vase with Design of Pines and Cranes (Modern Times)

在新石器时代至秦汉以前，岫玉是古玉器的主要原料。在红山文化、良渚文化，以及其他原始文化遗址中都有岫玉玉器出土；在河南安阳殷墟妇好墓出土的玉器中也发现了岫玉玉器；河北满城西汉中山靖王刘胜墓中出土的金缕玉衣也是用岫玉制成。岫玉是目前产量最高的玉料，中国每年出产的玉石原料中，岫玉占60%以上。

岫玉是地层中的超基性岩石经过漫长的岩浆腐蚀而逐渐形成的，其主要成分为镁质含水碳酸盐，此外还包括少量的透闪石、透辉石、方解石、白云石、水镁石等。岫玉是半透明至微透明的玉石，玉质细腻，性韧，较易打出断口，断口呈参差状，抛光后呈蜡状光泽。岫玉颜色多样，主色为豆绿色、黄绿色，其他颜色有青绿色、褐色、褐黄色、黄色、红色等。岫玉摩氏硬度为2.5～5.5度，密度为2.44～2.8g/cm^3，折射率为1.53～1.57。

不同产地的岫玉，矿物组成上会有所差异。根据产地不同，岫玉可分为辽宁岫玉、酒泉岫玉、昆仑岫玉、京黄岫玉、南方岫玉。辽宁

Xiu jade was originally an ultrabasic rock embedded in stratum which, after a long-term corrosion by lava, turned into jade. It is mainly composed of magnesian hydrous carbonate, as well as a small amount of tremolite, diopside, calcite, dolomite, brucite, etc. *Xiu* jade appears slightly transparent or semi-translucent and has delicate texture, tough yet pliable, and easily fractures with an uneven cross-section which has wax-like luster after polishing. Various in colors, *Xiu* jade is mainly in pea-green or yellow-green, as well as other colors including bluish green, brown, brownish yellow, yellow and red, etc., with a hardness of Mohs 2.5-5.5, density of 2.44-2.8 g/cm^3 and refraction index of 1.53-1.57.

Xiu jade mined in different areas varies in mineral composition, and based on its producing area, *Xiu* jade can be generally divided into five groups, including Liaoning *Xiu* jade, Jiuquan *Xiu* jade, Kunlun *Xiu* jade, Beijing Yellow *Xiu* jade and Southern *Xiu* jade. Liaoning *Xiu* jade is produced at Xiuyan County in Liaoning Province. It is mainly pea-green in color with delicate texture and high hardness. Jiuquan *Xiu* jade is produced at Qilian Mountains in Jiuquan City of Gansu Province, usually in dark-green

岫玉是指辽宁省岫岩县产出的岫玉，以豆绿色为主色，质地细腻，硬度高；酒泉岫玉是指产于甘肃省祁连山酒泉地区的岫玉，多呈墨绿色、黑色条带状，硬度较低；昆仑岫玉是指产于昆仑山的岫玉，质地细腻，多呈淡绿、淡黄、黄绿、暗绿、灰白、白色，有蜡状光泽；京黄岫玉是指产于北京市十三陵的老君堂的岫玉，呈淡黄至黄色；南方岫玉是指产于广东省信宜市的岫玉，呈黄绿色、绿色，有蜡状光泽。

tone and black stripped shape, lower in hardness. Kunlun *Xiu* jade refers to the jade mined at Kunlun Mountains. It is delicate in texture with wax-like luster and mostly has colors of light green, light yellow, yellowish green, dark green, gray and white. Beijing Yellow *Xiu* jade is those produced at Laojun Tang of the Ming Tombs in Beijing and varies from light yellow to yellow in color. Southern *Xiu* jade is mined at Xinyi County of Guangdong Province. It is mainly yellowish green or green in color and has wax-like luster.

• 岫玉"十八罗汉"山子（现代）
Xiu Jade *Shanzi* (Jade Display with Landscape Design) with Design of Eighteen Buddhist Arhats (Modern Times)

酒泉玉与夜光杯

　　唐朝诗人王翰有一首著名的《凉州词》中写道："葡萄美酒夜光杯，欲饮琵琶马上催。"夜光杯多由酒泉玉制成，具有耐高温、抗低温的优点。

　　早在西周之时，西域夜光杯就是贡品了。当时夜光杯由和田玉制成，直接运往长安、洛阳等地。后来由于玉杯在运输途中经常出现损坏的现象，所以就改为把和田玉运到酒泉，在那里再加工成夜光杯，然后再向东运输。后来因种种原因，和田玉不能及时供应，于是就改用酒泉玉来制作夜光杯。

Jiuquan Jade and *Yeguang* Cup (literally the Luminous Wine Glass)

Wanghan, a poet of the Tang Dynasty, once alluded to *Yeguang* Cup in his poem *A Song of Liangzhou (Liangzhou Ci)*, says that "holding the *Yeguang* Cup filled with vintage wine; however the departing *Pipa* song is played to urge me." Most *Yeguang* Cups are made of Jiuquan jade, a material with the features of thermostability and low temperature-resistance.

　　It is said as early as in the Western Zhou Dynasty (1046 B.C.-221 B.C.), *Yeguang* Cup of Western Regions (Xiyu, ancient name of nations and regions on the west of Yumen Pass and Yang Pass) had been presented as the tribute to the court. At that time, it was mainly made of Khotan jade, and then transported directly to Chang'an and Luoyang. Given that the jade cups would be damaged in long-distance transportation, so people delivered Khotan jade to Jiuquan City, made it into *Yeguang* Cup there and then carried the cups eastward. Later, due to various reasons, Khotan jade was short in supply and as a result, Jiuquan jade was used to carve the *Yeguang* Cup.

绿松石

　　绿松石又称"松石"。中国的绿松石产于湖北省襄阳地区的郧县、郧西县、竹山县。绿松石在中国古代使用广泛，历代文物中均有绿松石制品。在良渚文化遗址以及

Turquoise

Turquoise, or *Songshi*, is distributed in Yun County, Yunxi County and Zhushan County of Xiangyang City, Hubei Province. It was widely used in ancient China and turquoise articles were discovered in cultural relics of

商代墓葬中出土有绿松石饰品，颜色鲜艳，独具特色。

绿松石是含水的铜铝磷酸盐，此外还含埃洛石、高岭石、石英、云母等矿物。不少绿松石有球粒、环带、葡萄状构造，并有石英晶粒。比较软的松石，孔隙度较大，

- 绿松石串珠（商）
Turquoise Strung Beads (Shang Dynasty, 1600 B.C.-1046 B.C.)

- 绿松石耳坠（良渚文化）
Turquoise Ear Pendants (Liangzhu Culture, 4000-5000 years ago)

each dynasty. For instance, turquoise ornaments were excavated in the Liangzhu Culture site as well as in the tombs of the Shang Dynasty (1600 B.C.-1046 B.C.). They are bright in color and have distinctive features.

Turquoise is a hydrous phosphate of copper and aluminum and also contains minerals such as halloysite, kaolinite, quartz and mica, etc. Quite an amount of turquoise has spherulitic, cyclic and botryoidal structure, and containing quartz crystal particle. The softer turquoise is usually higher in porosity and has 18%-20% of adsorbed water and constitution water in the pores, therefore is liable to acid corrosion. It is opaque (the transparent turquoise slice is of the low quality which contains excess quartz) and easily fractures either in the process of carving or in storage, and the cross-section is neat and smooth, without any specific direction. Its color is bright, ranging from sea-blue to green, while its luster depends on its hardness and often the higher is the hardness, the brighter is the luster. In terms of hardness, turquoise generally falls into three levels. Porcelain turquoise (*Cisong*) takes the first place with a hardness of Mohs 5.3, and then

孔隙里含18%～20%的吸附水、结构水，易受酸的腐蚀。绿松石不透明（可透光的绿松石薄片是含石英量过大的次品）。绿松石在制作过程中或存放时常常自然裂开，裂开面平整光滑，方向性不强。绿松石颜色醒目，从海蓝色到绿色都有。其光泽会随硬度不同而变化，硬度高的绿松石光泽好，硬度低的绿松石光泽差。绿松石的摩氏硬度分为三个级别：最硬的是"瓷松"，摩氏硬度约5.3度；其次是硬松石，摩氏硬度4.5～5.3度；最软的"面松"摩氏硬度在4度以下。绿松石的密度为2.66～3.08 g/cm^3，折射率为1.60。

根据颜色、硬度和品质的不同，绿松石一般可分为瓷松、绿松、面松和铁线松。瓷松是质地最硬的绿松石，因抛光后的光泽、质感与瓷器相似而得名，颜色通常为纯正的天蓝色，是绿松石中的上品；绿松的硬度比瓷松略低，颜色从蓝绿到豆绿色，属中等品；面松又称"泡松"，颜色从淡蓝色到月白色，质地较软且疏松，可用小刀刻划，是品质最差的绿松石，只有大块者才具有使用价值；铁线松是

comes hard turquoise with a hardness of Mohs 4.5-5.3, and finally flour turquoise (*Miansong*) with hardness lower than Mohs 4. Its density and refractive index are respectively 2.66~3.08 g/cm^3 and 1.60.

Based on the differences in color, hardness and texture, turquoise is usually divided into four categories, porcelain turquoise (*Cisong*), green turquoise, flour turquoise (*Miansong*), and iron-wire turquoise. Porcelain turquoise (*Cisong*) derives its name from its luster and texture which are similar to those of porcelain after polishing. It is usually in pure sky-blue, and is the hardest and also the best in quality. Compared with porcelain turquoise (*Cisong*), green turquoise is slightly lower in hardness and average in quality with its color ranging from bluish green to pea-green. Flour turquoise (*Miansong*), also named bubble turquoise, varies from light blue to pale blue in color. Because of a soft and loose texture it can be carved by a small knife and therefore is lowest in quality. Only the large pieces possess use value. Iron-wire turquoise has black tortoise-shell pattern, web pattern or vein pattern across the surface. These veins,

指有黑色龟背纹、网纹或脉状纹的绿松石品种，铁线（绿松石的纹理）纤细，质地坚硬，与绿松石形成一体，其中以具有蜘蛛纹的绿松石为佳品。

known as iron-wire, are thin and fine, hard in texture and integrated with the turquoise as a whole. The first-class iron-wire turquoises are those with pattern of spider web.

- 绿松石花卉瓶（现代）
Turquoise Vase with Floral Design (Modern Times)

- 绿松石寿阳公主（现代）
Turquoise Display with Design of Princess Shouyang(Modern Times)

昆仑玉

昆仑玉，又称"青海玉"，产于昆仑山麓，矿点多分布在青海省格尔木市西南、青藏公路沿线一百余公里处的高原丘陵地区。昆仑玉的使用历史悠久，在新石器时代文化遗址中就发现有昆仑玉制品。近年来，昆仑玉因被选为2008年北

Kunlun Jade

Kunlun jade (also Qinghai jade) is mined in the hillsides of Kunlun Mountains and has its deposits mostly distributed in the plateau and hills one hundred meters away from Qinghai-Xizang Highway, to the southwest of Golmud City in Qinghai Province. Kunlun jade was long used in history which could be certified by

京奥运会奖牌制作材料而名噪一时，价格扶摇直上，成为玉中的后起之秀。

昆仑玉是在地壳深处高温高压条件下，经过几千年的结晶冷却过程形成的。其矿物成分以透闪石为主，含量一般在95%之上，并且含有铁、硒、锌、镁、铅、钾、钙、硅等二十几种对人体有益的微量元素。质地细腻温润，油性好，韧性好，可雕性强。昆仑玉一般呈半透明状，具有油脂光泽。摩氏硬度6～6.5度，密度为2.7～3.1g/cm³，折射率为1.61左右。

按颜色特征，昆仑玉可分为白玉、青白玉、青玉、烟青玉、翠青玉等。白玉是昆仑玉的主要品种，也是产量最大的品种，外观呈灰白色至蜡白色，透明度好于和田白玉，有少量甚至能达到羊脂白玉的品质；青白玉外观为灰绿色至青灰色、浅黄灰色等，颜色淡雅清爽，半透明度明显大于和田青白玉；青玉外观呈青灰至深灰绿色，色调较闷暗，质地优于和田青玉，适宜制作大中型摆件、器皿；烟青玉外观呈灰紫色至烟灰色，也有人称其为紫罗兰、

the Kunlun jade wares unearthed in the Neolithic cultural sites. Having been selected as the medal material for the 2008 Olympic Games in Beijing, Kunlun jade again comes to fame in recent years, rises rapidly in price, and has a promising future.

Kunlun jade is embedded deeply in the crust and takes shape under high temperature and pressure after millennia of crystallization and cooling procedure. It is mainly composed of tremolite, taking up above 95 percent in the content, besides which there are nearly twenty kinds of trace elements such as iron, selenium, zinc, magnesium, lead, potassium, calcium, silicon, etc., all good for health. Kunlun jade is delicate and mild in texture, with good oiliness and ductility, and therefore suitable for carving. It appears semi-translucent and bright as coagulated grease. Its hardness, density and refractive index are respectively 6-6.5, 2.7-3.1 g/cm³, and 1.61 or so.

Based on the variation of color, Kunlun jade is divided into several types, including white jade, greenish-white jade, gray jade, smoke-gray jade, and bright-gray jade, etc. White jade is the main variety and also in greatest output.

藕荷玉、乌青玉等，是昆仑玉的一个标志性品种；翠青玉为翠绿色的昆仑玉，颜色似嫩绿色翡翠，与青玉、碧玉的颜色有明显不同。

It varies from gray to pale-white in color, has a transparency higher than that of Khotan jade and sometimes a few even can match mutton-tallow white jade in quality. The color of greenish-white jade ranges from gray-green to blue-gray or light yellow-gray, with simple and elegant tone. The Kunlun greenish-white jade is also more semi-translucent than Khotan greenish-white jade. Gray jade, varying from blue-gray to dark gray-green, has a dark tone and a texture better than Khotan Gray jade, fit for carving large and medium-sized display or vessel. Smoke-gray jade, a representative kind of Kunlun jade, differs from gray-purple to smoke-gray in color and therefore is also called violet jade, *Ouhe* jade (*Ouhe* refers to a lotus-like pale pinkish purple), and black-blue jade. Bright-gray jade derives the name from its color, which resembles the tender green of jadeite, but is quite different from that of gray jade and green jade.

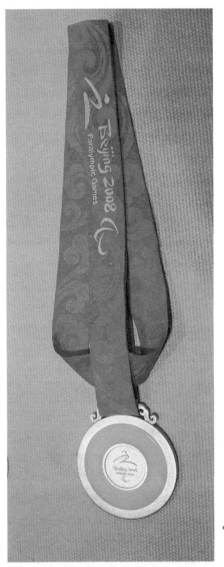

- 2008北京奥运会"金镶玉"奖牌 (图片提供：CFP)
  Gold-inlaid-with-jade Medal of 2008 Beijing Olympic Games

翡翠

　　翡翠，又称"缅甸玉"，是一种硬玉，主要产于缅甸。翡翠虽然主要产于缅甸，但其后期加工大部分是在中国完成，是中国玉器制作中常用的材料。在清代以前，和田玉被尊为"玉石之王"。但进入清代之后，翡翠开始取代和田玉而成为新的"玉石之王"。

Jadeite

Jadeite, also called "Burmese Jade", belongs to the chalchiguite family, mainly found in Burma. Nevertheless, its post-processing is largely completed in China, where jadeite is a commonly used material for producing jade articles. Before the Qing Dynasty (1616-1911), Khotan jade had been honored as "King of Jade", however, entering the Qing Dynasty, the position was replaced by jadeite.

　　Jadeite is a kind of jadeitite or omphacitite, with the equivalent commercial value of jade, whose composition is the silicate of sodium and aluminum, usually containing chromium, tantalum, iron, manganese, nickel and other inclusions. It displays an intertwined structure of filamentary crystal and porphyritic crystal, with visible spots, acicular, or flat glitter which is usually known as "jadeite nature". Jadeite often appears from semi-translucent to opaque, whose transparency depends on the size and arranging order of jadeite mineral particles. The higher its transparency is, the more valuable the jadeite will be. Jadeite has relatively

• 翡翠饕餮纹方鼎（清）

Jadeite *Ding* with *Taotie* (a Chinese Mythical Animal) Pattern (Qing Dynasty, 1616-1911)

翡翠是商业价值达到玉石级的硬玉岩或绿辉石岩，成分是含钠铝的硅酸盐矿物，常含铬、钽、铁、锰、镍等杂质元素。其结构是一种纤维状晶体与斑状晶体交织在一起的结构，并可见星点状、针状或片状闪光，这种闪光就是我们常说的"翠性"。翡翠一般呈半透明至不透明状，其透明度取决于翡翠矿物颗粒的大小和排列顺序等。透明度越高的翡翠价值越高。翡翠硬度和密度较高，摩氏硬度为6.5～7度，密度为3.30～3.36 g/cm³。

翡翠可根据"种""水""色""地"四个标准来划分种类。

"种"，是对翡翠质量综合评价的名称，即根据翡翠的质地、透明度，并参考绿色，作为划分翡翠的标准。人们把质地和透明度相

high hardness and density, whose Mohs hardness ranges from 6.5 to 7 and density ranges from 3.30 g/cm³ to 3.36 g/cm³.

Jadeite can be categorized according to four standards, which are "kind" (*Zhong*), "water" (*Shui*), "color" (*Se*) and "ground" (*Di*).

"Kind" (*Zhong*) is the term for an overall evaluation of jadeite's quality, i.e. the standard by which jadeite is divided according to its texture, transparency, and with reference to its green color. Jadeite materials, based on their texture and transparency, are divided into three kinds, specifically, old kind, new kind and new-old kind. Jadeite's kinds, be it "old" or "new", are not related with the time when jadeite materials come into being or when they are mined, but just a common expression. Old kind of jadeite is credited for its high-quality

● "老种"翡翠
Old Kind Jadeite

● "新种"翡翠
New Kind Jadeite

似的翡翠原料归为老种、新种、新老种三类。翡翠"种"的"老"和"新"不代表翡翠料形成时间或开采时间的早晚，只是一种习惯说法。老种翡翠是翡翠中材质优秀的原料，其矿物组成单一、矿物颗粒细小均匀，结构致密，质地细腻，绿色纯正浓艳，透明度好，硬度大；新种翡翠的矿物组成复杂，晶体颗粒粗糙，玉质松散，绿色杂而淡，透明度差，甚至不透明，硬度低，经济价值和收藏价值不大；新老种翡翠，是介于老种和新种之间的翡翠。

• "新老种"翡翠
New-old Kind Jadeite

"水"，又叫"水头"，表示翡翠的透明度，是影响翡翠品质的重要因素。通常用聚光电筒来观察翡翠，用光线照入的深浅来衡量水头的长短

material, which features single mineral composition, small and evenly distributed mineral particles, compact structure, fine texture, purely bright green color, high transparency and hardness. New kind of jadeite features complex mineral composition, rough crystal particles, loose texture, impurely light green color, low transparency (even to the extent of appearing opaque) and low hardness, which has limited economic value and collection value. New-old kind of jadeite features a quality in between those of the old kind and the new kind.

Water (*Shui*), also called "water head" (*Shuitou*), refers to jadeite's transparency, which is an important factor that affects jadeite's quality. Usually, jadeite is observed with a condenser flashlight by which, the water head's length and jadeite's transparency is measured by how far the beam of light can reach. Jadeite, with long "water head", features fine texture and evenly structured particles, which is of high value. Jadeite's mineral composition has an impact on its "water head". The purer the composition is, the longer the "water head" will be. Besides, the thickness of jadeite jewelry and the jadeite's color depth can, to some extent, affect the

和翡翠的透明度。"水头"长的翡翠质地细腻，颗粒结构均匀，价值高。翡翠矿物成分会影响"水头"，成分越单一，"水头"就越长。另外翡翠首饰的厚薄和自身颜色的深浅，在一定程度上也会影响"水头"，颜色浅、厚度薄者"水头"就显得长；反之，颜色深、厚度大者"水头"就显得短。

"色"，指翡翠的颜色。翡翠颜色多样，有商业价值的颜色是白、绿、红、紫、黄几种。

"地"，是评价翡翠材质的商业术语。以一件翡翠首饰而言，颜色漂亮的部位叫"旺色"，"旺色"之外的部位叫"地子"。全部是绿色的翠戒面石没有地子，也

"water head". Light colored and thinly cut jadeite's "water head" appears long, whereas dark colored and thickly cut jadeite's "water head" appears short.

"Color" (*Se*), referring to jadeite's color, is the most direct method for categorization. Jadeite is of various colors, among which, the ores of white, green, red, purple, and yellow have commercial values.

"Ground" (*Di*), is a business term for evaluating the materials of jadeite. For a piece of jadeite jewelry, the beautifully colored area is called the flourishing color (*Wangse*), while the rest is called the "ground" (*Dizi*). An all-green jadeite ring gem has no ground, or can be seen as a special kind of ground. According to its quality, jadeite's ground can be divided

• 红翡手镯
Red Jadeite Bracelet

• 金黄翡龙钩
Golden Jadeite Belt Hook with Dragon Design

可看成是一种特殊的地子。翡翠的地子按优劣可分为玻璃地、冰地、糯米地、豆地、芋头地、灰地等。其中，玻璃地是最好的地，地子明亮、制作成坠饰后会有猫眼状或圆状照映的闪光效果，中间有灰黑色光泽。

into glass ground, ice ground, sticky-rice ground, bean ground, taro ground, grey ground, and etc., among which, glass ground excels the others for its brightness, and has a cat's eye or circular glittering effect with grayish black luster in the center when crafted into pendants.

- "紫罗兰"翡翠手镯
  Violet Jadeite Bracelet

- "福禄寿"手镯
  "福禄寿"是指红、绿、紫三种颜色共存于一块翡翠上。
  "*Fu Lu Shou*" Bracelet
  "*Fu Lu Shou*" (respectively standing for "happiness", "fortune" and "longevity" in Chinese) refers to the co-existence of three colors, specifically, red, green, and purple, on one single piece of jadeite jewelry.

> 造型之美

中国玉器的造型多样，根据不同的用途，玉器可大致分为玉礼器、装饰玉、实用玉器、丧葬用玉和陈设用玉等几大类。

玉礼器

玉礼器在中国古代玉器中地位很高，因其象征着天命与王权而受到统治阶级的追捧。新石器时代良渚文化遗址中就有出土玉璧、玉琮等玉礼器。商周时，玉礼器的使用就已体系化了。《周礼》中将西周玉礼器分为三大类：第一类是以"六瑞"为代表的瑞玉；第二类是以"六器"为代表的祭器；第三类是朝廷使用的各种玉符节、玉兵器等。

随着时代的发展，玉礼器的品种发生了一些变化。有些玉礼器品种

> The Beauty of Shape

Various in shape, China's jade articles, according to their different applications, can be basically categorized into ritual objects, decorative articles, practical articles, funeral articles and furnishing articles.

Jade Ritual Objects

Jade ritual objects were highly positioned among ancient China's jade articles, and hotly pursued by the ruling class, for they symbolized providence and imperial power. From the Liangzhu Cultural Relics of Neolithic Age, jade *Bi*, jade *Cong* and other jade ritual objects have been excavated. During the Shang and Zhou dynasties (1600 B.C.-221 B.C.), jade ritual objects had been systemized. In the book *Rites of Zhou Dynasty (Zhouli)*, jade ritual objects of the Western Zhou Dynasty (1046 B.C.-771 B.C.) were divided into three kinds: firstly,

逐渐弃用，有些玉礼器演变为装饰用玉，也出现了一些新的玉礼器品种，如秦朝出现的玉玺、隋唐出现的玉带、清代宫廷盛行的玉如意等。

jade auspicious articles, represented by the Six Auspicious Items, secondly, sacrificial vessels, represented by the Six Ritual Objects, and thirdly, all kinds of jade tallies and weapons used by the imperial court.

With the advancement of time, jade ritual objects had undergone some changes in variety. Some kinds of jade ritual objects were gradually abandoned, while some were evolved into jade decorative articles. Also, some new kinds of ritual objects emerged, such as imperial jade seals (*Xi*) of the Qin Dynasty (221B.C.-206 B.C.), jade belts of the Sui and Tang dynasties (581-907), and jade *Ruyi* (furnishing articles that symbolize auspiciousness) that were prevalent in the Qing Dynasty (1616-1911).

Jade *Bi*: It is a flat jade disc with a circular hole in the center, which could be used for a long time and has many different kinds. Jade *Bi* emerged in the Neolithic Age (approx. 4500-8500 years ago), and had existed until the Qing Dynasty (1616-1911). During the Western Zhou Dynasty (1046 B.C.-771 B.C.), jade *Bi* was divided into blue *Bi* (*Cang Bi*), millet pattern *Bi* (*Gu Bi*) and cattail pattern *Bi* (*Pu Bi*). From the Warring States Period (475B.C.-221 B.C.) to the Western and Eastern Han dynasties (206 B.C.-220 A.D.), jade *Bi* had achieved its

- 重环谷纹玉璧（战国）
Double-ringed Jade *Bi* with Millet Pattern (Warring States Period, 475 B.C.-221 B.C.)

- 透雕双龙出廓玉璧（汉）
Openwork Jade *Bi* with *Chukuo* Design of Double-dragon (*Chukuo* refers to a jade sculpting, literally means cross the outline, indicating an advanced style with bulging openwork dragon, phoenix or tiger design carved along the rim of the jade ware, Han Dynasty, 206 B.C.-220 A.D.)

玉璧：中间有一圆孔的圆形板状玉器，使用时间长、品种数量多。玉璧自新石器时代出现一直延续到清代，西周时有苍璧、谷璧、蒲璧之分，在战国至两汉时期发展至顶峰，出现了镂雕出廓璧。璧最初是作为礼器出现，后来也作为装饰品、馈赠用品或随葬品使用。

heyday with the emergence of openwork *Chukuo Bi* (*Chukuo* refers to a jade sculpting, literally means cross the outline, indicating an advanced style with bulging openwork dragon, phoenix or tiger design carved along the rim of the jade ware). Originally appeared as a kind of ritual object, *Bi* was later used as a decorative article, gift or funerary object as well.

历代玉璧特点
Features of Jade *Bi* During Different Dynasties

年代 Period	特点 Features	图片 Picture
新石器时代 Neolithic Age (4500-8500 years ago)	形状大而厚，素面为主，玉质不佳，多呈青灰色、灰白色。 Big and thick in shape, primarily plain in appearance; low-quality in jade texture, mostly caesious and grayish white in color.	 • 玉璧（良渚文化） Jade *Bi* (Liangzhu Culture, 4000-5000 years ago)
夏商周 Xia, Shang, and Zhou Dynasties (approx. 2070 B.C.-221 B.C.)	大小不等，数量增多，多以云纹、龙凤纹、鸟纹、兽面纹装饰，以青白玉为主。 Various in size; increasing in quantity; mostly decorated with cloud pattern, dragon-phoenix pattern, bird pattern, animal-faced pattern, dominated by greenish-white jade.	 • 龙纹玉璧（西周） Jade *Bi* with Dragon Pattern (Western Zhou Dynasty, 1046 B.C.-771 B.C.)

春秋战国至汉代 Spring and Autumn Period and the Warring States Period to the Han Dynasty (770 B.C.-220 A.D.)	大小不等，出现了出廓璧（圆形玉璧外有突出的部分），八角形、菱形、椭圆形等形状的玉璧。纹饰增多，工艺更加精细，以和田玉为主要材质。 Various in size; having emerged *Chukuo Bi* (with bulging openwork carving along the outline of the *Bi*), and octagonal shaped, rhombic shaped, elliptical shaped jade *Bi*; increasingly rich in pattern and design, more delicate in craftsmanship, with Khotan jade being the main material.	 • 双凤谷纹璧（战国） Jade *Bi* with Millet Pattern and Double-phoenix Design (Warring States Period, 475 B.C.-221 B.C.)
唐至清代 Tang Dynasty to the Qing Dynasty (618-1911)	形制和纹饰多仿古，雕刻精细，纹饰生动形象，特别是多层透雕，立体感强，以和田玉为主要材质。 Modeling on antique jade *Bi* in terms of shape, pattern and design; exquisite in carving, vivid in pattern and design, especially notable for the multi-layered openwork carving which looked stereoscopic, with Khotan jade being the main material.	 • 云龙纹玉璧（唐） Jade *Bi* with Cloud-and-dragon Pattern (Tang Dynasty, 618-907)

玉圭：长方形的扁平玉器，分为平首圭和尖首圭。圭是国君所用的瑞器，代表君权神授，是执掌国家最高权力的象征，因而是最重要的玉礼器。西周时圭有十几种，有镇圭、桓圭、信圭、躬圭、谷圭、土圭、珍圭、琬圭、琰圭等，都有不同的用途。汉代以后玉圭就很少见，明清时期又出现仿古玉圭。

Jade *Gui*: a rectangular shaped tablet of jade, which is either sharp topped or flat topped. *Gui* is a kind of auspicious item used by emperors, which represented the divine rights of emperors and symbolized the highest authority that governed the country, and therefore was the most important jade ritual object. During the Western Zhou Dynasty, *Gui* had more than ten kinds for different applications, including *Zhen Gui*, *Huan Gui*, *Xin Gui*, *Gong Gui*, *Gu Gui*, *Tu Gui*, *Zhen Gui*, *Wan Gui*, *Yan Gui* and etc. After the Han Dynasty (206 B.C.-220 A.D.), jade *Gui* had been rarely seen until the Ming and Qing dynasties when imitation jade *Gui* emerged.

- 尖首玉圭（商）
Jade *Gui* with Sharp Top (Shang Dynasty, 1600 B.C.-1046 B.C.)

- 仿古平首玉圭（清）
Imitation Jade *Gui* with Flat Top (Qing Dynasty, 1616-1911)

玉琮：中心有一贯通上下的方形柱体玉器，形制多样，器形有高有矮。多用于礼地、祭神灵、殓尸等场合。玉琮在新石器时代良渚文化和齐家文化遗址中大量出土，商代时达到鼎盛，汉代以后就消失了，宋以后出现仿古玉琮。

Jade *Cong*: It is an octagonal shaped jade article with a thorough square cylinder in the center, which could be various in shape and height, often used for earth worshiping, sacrificial and coffining occasions. A large number of jade *Cong* were excavated from the Liangzhu Culture and Qijia Culture Relics. Jade *Cong* had its heyday during the Shang Dynasty (1600 B.C.-1046 B.C.) and disappeared after the Han Dynasty (206 B.C.-220 A.D.). Since the Song Dynasty (960-1279), imitation jade *Cong* has appeared.

Jade *Zhang*: It is a rectangular shaped tablet of jade with an oblique

• 高型玉琮（良渚文化）
High-moulded Jade *Cong* (Liangzhu Culture, 4000-5000 years ago)

• 矮型玉琮（良渚文化）
Short-moulded Jade *Cong* (Liangzhu Culture, 4000-5000 years ago)

玉璋：形似玉圭，长方形体，扁平状，一端为斜刃，另一端有穿孔。西周时玉璋有五种：赤璋、大璋、中璋、边璋、牙璋，分别具有不同的用途。其中牙璋属节符器，是调动军队的凭证，是一种很重要的玉节（西周天子派人传达命令时使用的信物）。

blade on one end and a hole on the other, resembling jade *Gui* in appearance. During the Western Zhou Dynasty (1046 B.C.-771 B.C.), there were five kinds of jade *Zhang* for different applications, specifically, red *Zhang* (*Chi Zhang*), big *Zhang* (*Da Zhang*), medium *Zhang* (*Zhong Zhang*), and *Zhang* with teeth-shaped design (*Ya Zhang*). *Ya Zhang* was a kind of tally, a credential to mobilize the troops, which was a very important jade tally used by the Son of Heaven

- 玉璋（商）
  Jade *Zhang* (Shang Dynasty, 1600 B.C.-1046 B.C.)

- 墨玉牙璋（新石器时代）
  Black Jade *Zhang* with Teeth-shaped Design (the Neolithic Age, 4500-8500 years ago)

玉璜：圆弧形玉器，一般都说"半璧曰璜"，因为很多玉璜是用玉璧的残件加工改制而成的，但实际上很少有标准的半璧形的玉璜，多数璜不及半璧，也有的超过半璧。玉璜除作礼器使用之外，也作为佩饰，尤其是作为玉组佩中的组成部分大量存在。春秋时期出现了钻孔的玉璜，战国、汉代时期还出现有镂空璜、连璜、附加纹饰璜、龙首璜等。

(*Tianzi*) of the Western Zhou Dynasty when sending a herald to deliver an order.

Jade *Huang*: It is an arc-shaped jade article, with the common saying of *Huang* is half *Bi*, because a large number of jade *Huang* were processed with defective jade *Bi*, however, in reality, there were seldom semi-*Bi* shaped jade *Huang*. The majority of *Huang* were less than half *Bi* in appearance, while some were more than it. Except for being used as ritual objects, jade *Huang* were also used as accessories, especially as a component of a set of jade pendants. Jade *Huang* with a hole emerged in the Spring and Autumn Period (770 B.C.-476 B.C.), and openwork *Huang*, connected *Huang* (*Lian Huang*), *Huang* with additional pattern and *Huang* with dragon-headed design emerged in the Warring States Period (475 B.C.-221 B.C.), the Western Han and Eastern Han dynasties (206 B.C.-220 A.D.).

Jade *Hu*: It is a tiger-shaped tablet of jade. According to the book *Rites*

- 双龙首玉璜（汉）
  Jade *Huang* with Double-dragon-headed Design (Han Dynasty, 206 B.C.-220 A.D.)

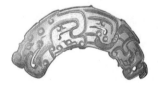

- 镂雕虎纹玉璜（西周）
  Openwork Jade *Huang* with Tiger Pattern (the Western Zhou Dynasty, 1046 B.C.-771 B.C.)

玉琥：扁平片状的虎形玉件。据《周礼》记载，玉琥是"礼西方"和"祈雨"的玉礼器，但从出土的玉琥情况来看，大多应属装饰品和艺术品。还有一种说法是玉琥是作为发兵的凭证——虎符来使用。

- 玉琥（商）

Jade *Hu* (Shang Dynasty, 1600 B.C.-1046 B.C.)

玉符节：朝廷重要的信物。玉符的形制特点是把一件器物剖分为两半，朝廷派官员驻守外地，便把这符的一半交他带去，以后如果朝廷有命令下达，去传达命令的人就要拿着符的另一半，作为信证，驻外官员把两片一合，能组合成一个完整的，便确认所传命令确实是朝廷所发。玉符有虎符、鱼符、龟符、麟符、鹤符等。

玉兵器：主要有玉刀、玉戈、玉钺、玉戚等，器形与同名石器或

of Zhou Dynasty (*Zhouli*, one of three ancient ritual books listed among the classics of Confucianism), jade *Hu* was used as a jade ritual object to worship the West and pray for good rain. However, most unearthed jade *Hu* were used as ornaments and artworks. Besides, some say that jade *Hu* was used as Tiger Tally (a tiger-shaped tally issued to generals as imperial authorization for troop movement in ancient China), which was a credential to dispatch the troops.

Jade Tally (*Fujie*): It is an important credential of the imperial court, which was divided into two halves. One half would be given to the official, who was sent to station an army in a local place. If the imperial court issued an order to him, the herald would take the other half as a credential. The local official would piece together the two halves to complete a whole for verification. Jade tallies included Tiger Tally (*Hufu*), Fish Tally (*Yufu*), Turtle Tally (*Guifu*), Unicorn Tally (*Linfu*), Crane Tally (*Hefu*) and etc.

Jade Weapons: It mainly refers to jade knife, jade *Ge* (sickle-shaped weapon), jade *Yue* (axe-shaped weapon,

青铜器相同，虽然有兵器之形，但出土的玉兵器大多没有使用过的痕迹，应属礼仪用器，广泛地用于祭祀、仪仗活动。

turning into ritual object afterwards), jade *Qi* (axe-shaped ritual object, usually with a big punched hole in the center), and other kinds of jade weapons, all shaped like stone or bronze weapons with the same names. Although appearing like weapons, most unearthed jade weapons bear no traces of wear and tear, which might have been ritual objects, widely used for sacrificial and ceremonial occasions.

- 仿古玉虎符（清）
  Imitation of Jade Tiger Tally (Qing Dynasty, 1616-1911)

玉兵器的种类
Different Kinds of Jade Weapons

玉刀：由新石器时代的石刀演变而来，其器形主要有两种：一种为扁平状的长方形，一侧为刀背，一侧为刀刃；另一种是做成了带柄的形状。有的玉刀没有柄，只在刀身上有穿孔。

Jade Knife: It evolving from stone knives in the Neolithic Age (4500-8500 years ago), had two kinds of shapes. One featured a rectangular shaped tablet with blade on one side and the back of blade on the other. The other was shaped with a hilt. Some jade knives didn't have a hilt, but only bore a hole on their knives-blades.

- 玉刀（商）
Jade Knife (Shang Dynasty, 1600 B.C.-1046 B.C.)

- 四孔玉刀（龙山文化）
Jade Knife with Four Holes (Longshan Culture, 2500 B.C.-1900 B.C.)

玉戈：形状似刀，由"援"（刃部）和"内"（似柄有孔能穿系）两部分组成。玉戈出现于商代，西周时已作为象征权力的礼器使用。特别小的玉戈还可能充当货币，或作为装饰品。

Jade *Ge* (sickle-shaped weapon), resembled knife in appearance, constituted of two parts: *Yuan* (blade) and *Nei* (hilt-like part with a hole on it for attachment). Emerged in the Shang Dynasty (1600 B.C.-1046 B.C.), jade spears had been used as ritual objects that symbolized power during the Western Zhou Dynasty (1046 B.C.-771 B.C.). Tiny jade spears could be used as currency or ornaments as well.

- 玉戈（商）
Jade *Ge* (Shang Dynasty, 1600 B.C.-1046 B.C.)

- 玉钺（龙山文化）
Jade *Yue* (Longshan Culture, 2500 B.C.-1900 B.C.)

玉钺：由石斧演变而来，多为扁宽片状，有两面对削的弧形刃，近肩处有穿孔，制作精美的玉钺为军事首领的标志物，大型玉钺后来成为王权的象征。

Jade *Yue*: Evolving from stone axes, mainly featured a broad tablet with an arc-shaped blade formed by being sharpened from both sides, and had holes bored near the shoulder. A delicately crafted jade *Yue* was a mark of a military leader. Largely shaped jade *Yue* later became a symbol for the imperial power.

玉戚：一种形状类似石斧的兵器，中心有穿孔，两侧均有锯齿状，形状有两种，一种略呈长方形，刃部略弧；另一种背部稍圆，两侧近直，刃分4段，两面直刃，中间钻一大圆孔，又称璧戚。玉戚主要是王公贵族们在举行重大典礼时作仪仗器使用。

Jade *Qi*: It refers to an axe-like weapon with a hole in the center, jaggedly shaped on both sides, which had two kinds of shapes. One appeared rectangular, with a slightly arc-shaped blade. The other appeared slightly round on the back and straight on both sides, whose blade was divided into four parts, also called *Bi-Qi* with straight blades on both sides and a round hole in the center. Jade *Qi* was mainly used as ceremonial articles during great ceremonies held by kings and aristocrats.

玉戚（西周）
Jade *Qi* (the Western Zhou Dynasty, 1046 B.C.-771 B.C.)

装饰用玉

Jade Decorative Articles

装饰用玉是玉器中种类与数量最多的一类，包括玉佩饰、玉剑饰、玉服饰、玉首饰等。装饰玉不仅满足了人们对美的追求，更显示了佩戴者身份的尊贵，在中国玉器中占有相当重要的地位。

玉佩饰：指随身佩戴的装饰用玉，形制不一，但有个共同特点，就是都有孔，用于系绳佩带。常见的玉佩饰有玉觿、环玦、玉组佩、

Jade decorative articles took up a large percentage of jade articles, most diverse in kinds, including jade accessories, jade sword fittings, jade trappings, jade jewelry and etc. Jade decorative articles, not only fulfilled people's desire to pursue beauty, but also manifested the distinguished statuses of the wearers, highly positioned among China's jade articles.

Jade Accessories: It refers to jade decorative articles worn by people, various in shape, but all bore a hole on

玉勒、玉牌、玉刚卯等。有的玉佩饰还兼具实用功能，如玉觿、玉勒分别用于解结、束带之用。

them for attaching strings. Commonly, jade accessories included jade *Xi*, *Huan Jue*, a set of jade pendants, jade *Le*, jade plates, jade *Gangmao* and etc. Some jade accessories also had practical applications. For example, jade *Xi* and jade *Le* could be used to undo a knot, and bind up respectively.

玉佩饰的种类
Different Kinds of Jade Accessories

玉觿：觿是古人用骨头仿制兽牙形的装饰品，基本形状为一头宽厚，一头尖锐。最初的觿是用牛羊角制作，玉觿的制作是为了迎合人们佩戴觿的风俗需要。玉觿由于一端比较尖细，所以具有解结的功能。

Jade *Xi*: It was made by ancient people in imitation of beast tooth for decoration, basically shaped with one end broad and thick and the other sharp-pointed. Originally, *Xi* was made with horns of cattle and sheep. Jade *Xi* was crafted to meet people's customary need for wearing *Xi*. Since it was sharp-pointed on one end, *Xi* could be used to undo a knot.

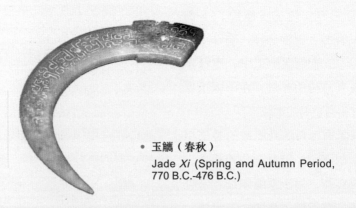

玉觿（春秋）
Jade *Xi* (Spring and Autumn Period, 770 B.C.-476 B.C.)

玉环：中间有大孔的圆形玉器。玉环起源于新石器时代，到战国时期达到顶峰，出土了许多花纹精致的玉环。唐宋以后，玉镯开始出现并取代了玉环的位置。

Jade Ring: It refers to a circular shaped jade article with a big hole in the center. Originated in the Neolithic Age (4500-8500 years ago), jade rings achieved its heyday during the Warring States Period, to which period many unearthed jade rings with exquisite patterns can be dated back. Since the Tang and Song dynasties, jade Bracelets have emerged and replaced jade rings as ornaments.

- 云雷纹玉环（战国）
Jade Ring with Thundercloud Pattern (Warring States Period, 475 B.C.-221 B.C.)

玉玦：有缺口的环形玉器。玉玦可作随葬殓尸时的耳饰，也可作佩于腰间的饰品。还有一个特别的功能，就是利用"玦"与"决"的谐音作为绝交的信物。

Jade *Jue* refers to a circular shaped jade article with an opening on it, which could be used as earrings during coffining occasions, or as an ornament hung from one's waist. It could also specially function as a pledge of splitting up with a friend, for *Jue* is homophonic with another Chinese character which means splitting up.

- 龙纹玉玦（西周）
Jade *Jue* with Dragon Pattern (Western Zhou Dynasty, 1046 B.C.-771 B.C.)

玉组佩：由璜、珩、环、冲牙及其他佩玉组成的玉佩饰，一般佩戴于胸前或腰间。佩戴者行走时，各种玉饰锵然相击，发出清脆悦耳之声。

Pendant Set: It was constituted of *Huang*, *Heng*, *Huan*, *Chongya*, and other kinds of jade pendants, often hung from one's neck or one's waist. When the wearer walked, all kinds of jade pendants would clash with each other, emitting clear and melodious sounds.

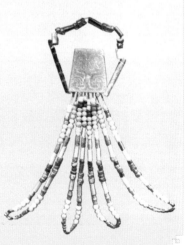

- 玉组佩（西周）
Jade Pendant Set (Western Zhou Dynasty, 1046 B.C.-771 B.C.)

玉佩：用薄片玉板制成的单件佩玉，形制多样，有动物形玉佩、花形玉佩、鸡心玉佩等。玉佩不仅仅是简单的装饰品，还具有表明身份、地位的功能。不同身份的人，佩玉的色泽、质地等都有区别。

Jade Pendant: It refers to a single piece of jade accessories made of jade tablet, which could be crafted into various shapes, such as animal-shaped, flower-shaped, and heart-shaped pendants. Jade pendants were not only simple ornaments, but also used to manifest the wearers' statuses and positions. The jade pendants worn by people with different statuses differed in color and texture.

• 玉虎佩（西周）
Jade Tiger Pendant (Western Zhou Dynasty, 1046 B.C.-771 B.C.)

• 折枝花玉佩（金）
Jade Pendant with Folding-branch Floral Pattern (Jin Dynasty, 1115-1234)

玉牌：方形或长方形的片状佩玉。玉器表面多有图案与文字，均浅浮雕或镂空雕刻，玉牌中间有孔作穿绳佩系之用。

Jade Tablet: It refers to a square or rectangular shaped tablet of jade. This kind of jade article's surface was often crafted with patterns and words, which were either shallow-relief or openwork carving. There was a hole in the center of a jade plate for attaching strings.

• 山水纹玉牌（明）
Jade Plate with Landscape Pattern (Ming Dynasty, 1368-1644)

玉刚卯：一种挂在腰上的护身符，上面刻有文字，为汉代特有的佩饰。一般呈方柱体，从上到下贯穿一孔，可以穿绳佩系。

Jade *Gangmao*: *Gangmao* refers to a kind of periapt hung from one's waist with words engraved on it, which was a distinctive kind of pendants, exclusively crafted in the Han Dynasty (206 B.C.-220 A.D.). It often featured a square column with a hole on it penetrating from top to bottom for attaching strings.

• 白玉刚卯（汉）
White Jade *Gangmao* (Han Dynasty, 206 B.C.-220 A.D.)

玉剑饰：用于剑上的装饰玉件，分别由剑首、剑格、剑璏、剑珌四件玉饰物组成。玉剑饰最早出现在西周，到春秋战国时期成为达官贵族标榜财富实力、显示社会地位、表明身份的象征。

Jade Sword Ornaments: It used as jade decorative articles on a sword, were constituted of four jade ornaments, which were sword pommel (*Shou*), sword guard (*Ge*), sheath buckle (*Zhi*) and sheath chape (*Bi*). Emerging in the Western Zhou Dynasty (1046 B.C.-771 B.C.), jade sword fittings became the symbol for wealth, social status and positions of aristocrats during the Spring and Autumn Period and the Warring States Period (770 B.C.-221 B.C.).

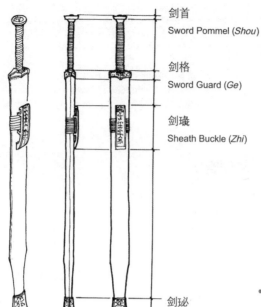

• 玉剑饰位置示意图
Illustration on the Positions of Jade Sword Fittings

玉剑首是剑柄端部的玉饰，常见的有圆形和方形两种。

Jade sword pommel (*Shou*) was a jade ornament on the top of the sword hilt, commonly circular shaped or square shaped.

• 玉剑首(西汉)
Jade Sword Pommel (Western Han Dynasty, 206 B.C.-25 A.D.)

玉剑格是剑柄与剑身之间的玉饰，扁长形，中部向两面凸起，两端薄，截面为菱形，中间有穿孔。

Jade sword guard (*Ge*) was a jade ornament positioned between the sword hilt and the sword body, elongated in shape, convex at the center, thin at both ends, with a hole in the center, whose cross-section appeared rhomboidal.

- 白玉浮雕螭纹剑格(西汉)
  White Jade Sword Guard (*Ge*) with *Chi*-dragon Design in Relief (Western Han Dynasty, 206 B.C.-25 A.D.)

玉剑璏是剑柄与剑锋之间的玉饰，镶嵌在剑鞘上，体积较大。

Jade Sheath buckle (*Zhi*) was a jade ornament positioned between the sword hilt and the sword edge, inlaid into the sword Sheath, large in volume.

- 双螭纹玉剑璏(西汉)
  Jade Sheath Buckle (*Zhi*) with Double-*Chi*-dragon Pattern (Western Han Dynasty, 206 B.C.-25 A.D.)

玉剑珌是剑鞘下端的玉饰，通常为扁长的梯形，用于保护剑鞘。

Jade Sheath chape (*Bi*) was a jade ornament at the bottom of the Sheath, usually elongated and trapezoid shaped, used to protect the Sheath.

- 象首纹玉剑珌(春秋)
  Jade Sheath Chape (*Bi*) with Elephant Head Pattern (Spring and Autumn Period, 770 B.C.-476 B.C.)

玉服饰：与服装配套使用的玉饰件，常见的有帽正、翎管、帽顶、玉带板、玉带钩、玉带扣、玉提携等，均有特定的用途。

Jade Trappings: They refer to jade ornaments used in conjunction with costumes, commonly including hat adjustment (*Maozheng*), feather tube (*Lingguan*), hat top (*Maoding*), jade belt tablet, jade belt hook, jade belt buckle, jade *Tixie* and etc, which had a particular application respectively.

玉服饰的种类
Different Kinds of Jade Trappings

玉帽正：缝在帽子上的装饰玉，起到"正冠"的作用，避免将帽子戴偏。一般为扁圆形、圆形、菱形片状，素面居多，也有的刻有纹饰。

Jade Hat Adjustment (*Maozheng*): It refers to a jade ornament sewed onto one's hat, which could be used to examine whether one's hat was rightly wore, usually featuring an oblate, circular, or rhomboidal shaped tablet, primarily plain in appearance. Some were carved with patterns as well.

- 白玉镂刻梅花纹帽正(明)
White Jade Hat Adjustment (*Maozheng*) with Openwork Plum-Blossom Design (Ming Dynasty, 1368-1644)

玉翎管：清代朝服官帽顶上插翎枝用的饰物，翎枝是清代官级的标志。玉翎管的形制为一端略粗的圆柱形，顶部有一突出的鼻，鼻上钻有一个连缀帽顶的横孔，粗的一端中空，用来插翎枝。

Jade Feathers Tube (*Lingguan*): It refers to an ornament on the top of an official's hat which was part of the Qing Dynasty's court dress, used to hold feathers (*Lingzi*), an indicator of one's rank among the Qing Dynasty's official system. Jade feathers tube was cylinder shaped with one end slightly broad. On top of it was a protruding nose with a flat hole on it for attaching feathers tube to the hat's top with a string. The other end was hollow, where a feather was inserted.

- 翡翠翎管（清）
  Jadeite Feathers Tube (*Lingguan*) (Qing Dynasty, 1616-1911)

玉帽顶：金元时期北方少数民族用于束发的玉器。玉帽顶采用钻孔透雕之法，内部镂雕有树木、草茎等纹饰，具有鲜明的时代特色。清代官员朝冠上的帽顶常以金为座，其上镶嵌珠宝、玉石等。

Jade Hat Top (*Maoding*): It refers to a jade article used by the northern ethnic groups to bind up their hair during the Jin and Yuan dynasties. Jade hat top was crafted with the technique of drilling and openwork carving, having trees, grass stems and other openwork patterns in the interior, displaying a distinctive feature of its period. The hat top on an official's hat of the Qing Dynasty was often made with a gold base and inlaid with bijou and jade.

- 金嵌珠宝帽顶（清）
  Gold Hat Top (*Maoding*) with Inlaid Gemstones (Qing Dynasty, 1616-1911)

玉朝珠：与清代朝服、公服配套使用的一种由108颗圆珠串成的项饰，由珊瑚、翡翠、玛瑙、水晶、琥珀、蜜蜡、碧玺等玉料制成。朝珠是皇权的象征物。按照清朝服饰制度，上到王公贵族、下到宫廷侍卫，一律佩戴朝珠。

Jade Court Beads (*Chaozhu*): It refers to a necklace strung with 108 beads, worn in conjunction with the court dress and official dress, crafted with coral, jadeite, agate, crystal, amber, honey amber, tourmaline and other jade materials. Court beads were a symbol for imperial power. According to the dress regulation of the Qing Dynasty (1616-1911), from kings and aristocrats to court guards, all should wear court beads.

• 翡翠朝珠（清）

Jadeite Court Beads (*Chaozhu*)(Qing Dynasty, 1616-1911)

玉带板：指由銙和铊尾组成的镶嵌在大带上的玉板。唐代官服实行大带制，以带上的饰物作为区别官职高低的标志。明代玉带板的使用更加普遍，形状和图案也更加多样。

Jade Belt Tablet: It refers to jade tablet that was inlaid into a broad belt, constituted of *Kua* and *Tuowei* (two kinds of ornament on a belt). The Tang Dynasty (618-907), implemented a Broad Belt system, which means that ornaments on one's belt were used as indicators for the ranks of officials. During the Ming Dynasty (1368-1644), jade belt tablets were more widely used, and more various in shape and pattern.

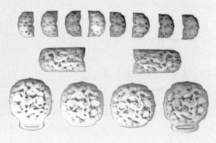

• 汪兴祖墓出土玉带板（明）

Unearthed Jade Belt Tablet from Wang Xingzu (Military Leader in the Ming Dynasty) Tomb (Ming Dynasty, 1368-1644)

玉带钩：春秋时期我国北方少数民族发明的勾连腰带的用具，由玉钩首、玉钩体、玉钩纽三部分组成，形制是一端曲首，背有圆纽。常见的玉带钩样式有棒形、竹节形、琴面形、圆形和兽形等。

Jade Belt Hook: It refers to a device for hooking up one's belt, invented by the northern ethnic groups during the Spring and Autumn Period (770 B.C.-476 B.C.), constituted of three parts, specifically, jade hook head, jade hook body and jade hook button. The head of it was curved and a round button was crafted on the back. Usually, jade belt hooks were stick shaped, bamboo-joint shaped, *Guqin* (an ancient stringed instrument)-surface shaped, circular shaped, animal-shaped and etc.

- 青白玉龙首螭纹带钩（元）
Greenish-White Jade Belt Hook with Dragon-headed and *Chi*-dragon Design (Yuan Dynasty, 1206-1368)

- 青玉马首带钩（清）
Gray Jade Belt Hook with Horse-headed Design (Qing Dynasty, 1616-1911)

玉带扣：一种束腰带用具。一般由两个方形或椭圆形的玉件组成，一件是钩，另一件是扣环，每件分为上下两层，上层有浮雕或镂雕纹饰，下层雕有穿带环。

Jade Belt Hook: It refers to a device for fastening the ends of a belt together. It was usually constituted of two square shaped or elliptic shaped jade articles. One was a clasp, and the other was a loop. Each had two layers. There were relief or openwork patterns on the upper layer, and a belt loop carved on the lower layer.

- 白玉龙首螭虎纹带扣（明）
White Jade Belt Hook with Dragon-headed *Chi*-and-tiger Pattern (Ming Dynasty, 1368-1644)

- 白玉镂雕螭虎纹带扣（明）
White Jade Belt Hook with Openwork *Chi*-and-tiger Pattern (Ming Dynasty, 1368-1644)

• 玉提携
Jade *Tixie*

玉提携：又叫"玉束带"，提携两侧均有一个扁形通孔，方便革带穿过。玉提携形制多样，有长方形、椭圆形、花形、荷叶形等。有些玉提携下部带有长而窄的玉环，作为悬挂物件之用。

Jade *Tixie*: It also called jade band, had a flat opening on each side for the convenience of passing a leather belt, various in shape, which could be rectangular shaped, elliptic shaped, flower shaped, lotus-leaf shaped and etc. Some jade *Tixie* had long and narrow jade rings at the bottom of them, used to hang other objects.

玉首饰：指头、颈、手部所佩戴的玉饰品。其中用于头部的玉饰品种类最多，有发饰、冠饰、耳饰等。发饰有玉梳、玉簪、玉发箍、玉戚璧等；冠饰主要是指与服饰配

Jade Jewelry: It refers to jade ornaments worn on one's head, around one's neck, and on one's hand. Jade head ornaments were most various in kind, including hair ornaments, hat ornaments, earrings, and etc. Hair ornaments included jade combs, jade hairpins, jade barrettes, jade *Qi-Bi* and etc. Hat ornaments mainly refer to hat adjustment (*Maozheng*), hat top (*Maoding*), feathers tube (*Lingguan*) and other jade ornaments used in full set. Earrings included jade *Jue*, jade *Tian* (hat ornament, hanging on the two sides of hat, used as earplug)

套使用的帽正、帽顶、翎管等；耳饰有玉玦、玉瑱等。用于颈部的玉饰品有项链、璎珞、佛珠、玉锁等，用于手部的则有手镯、扳指等。

and etc. Jade neck ornaments included necklaces, beaded necklace (*Yingluo*), Buddha beads, jade locks and etc. Hand ornaments included bracelets, thumb ring (*Banzhi*) and etc.

玉首饰的种类
Different Kinds of Jade Jewelry

玉梳：最初是礼器和身份的象征，后因中国古代女子有插梳为饰的习俗而逐渐成为一种装饰品。

Jade Comb: It originally a kind of ritual object and a symbol for status, gradually became an ornament, because ancient China's women followed the custom of fastening a comb onto their hair as an ornament.

- 玉梳（春秋）
  Jade Comb (Spring and Autumn Period, 770 B.C.-476 B.C.)

- 花卉纹玉梳背（唐）
  Jade Comb's Back with Floral Pattern (Tang Dynasty, 618-907)

玉簪：用来插定发髻或将冠固定在发髻上的玉饰。中国古代男女皆留长发，所以簪的使用很普遍。但玉簪比较珍贵，多为达官贵族或宫廷贵妇使用。

Jade *Zan* (Hairpin): Jade *Zan* refers to a jade ornament used to fasten one's topknot or fix a *Guan* (a special head-dress) onto one's topknot. Since both men and women in ancient China had long hair, *Zan* were widely used. However, jade *Zan* were relatively precious, mainly worn by aristocrats or court ladies.

- 玉簪（春秋）
  Jade *Zan* (Hairpin)(Spring and Autumn Period, 770 B.C.-476 B.C.)

- 玉簪（明）
  Jade *Zan* (Hairpin)(Ming Dynasty, 1368-1644)

- 凤首簪（唐）
  Zan (Hairpin) with Phoenix-headed Design (Tang Dynasty, 618-907)

玉镯：一种戴在手上的环形玉饰，截面有圆形、半圆形、方形、长方形等。有的素身无纹，有的刻有精美纹饰。明清时期出现了翡翠镯子，沿袭至今，深受女士喜爱。

Jade Bracelet: It refers to a ring-shaped jade ornament worn around one's wrist, whose cross-section could appear circular, semi-circular, square, or rectangular in shape. Some were plain in appearance without patterns, while some were carved with exquisite patterns. Emerging in the Ming and Qing dynasties, jadeite bracelets have existed until today, highly welcomed by ladies.

- 玉镯（良渚文化）
  Jade Bracelet (Liangzhu Culture, 4000-5000 years ago)

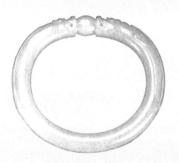

- 白玉二龙戏珠镯（清）
White Jade Bracelet with Two-dragon-playing-a-pearl Design (Qing Dynasty, 1616-1911)

- 翡翠手镯（现代）
Jadeite Bracelet (Modern Times)

项链：挂在颈上的装饰物，由较小的珠子或者管状物穿成一串。长度有的刚好绕颈一周，叫"卡脖链"，有的长可及胸，也有的长到可以绕颈二三周。清代以后的项链出现了"项坠"和"搭扣"，成为项链的重要组成部分。

Necklace: It refers to a kind of ornament hung around one's neck, strung with tiny beads or tube-shaped items. Some could be wound round one's neck exactly by one circle, called Tight necklace; some could reach the chest; some were as long as to be wound around one's neck by two or three circles. Since the Qing Dynasty (1616-1911), pendants and clasps have appeared, becoming important parts of necklaces.

- 玉项饰（西周）
Jade Neck Ornament (Western Zhou Dynasty, 1046 B.C.-771 B.C.)

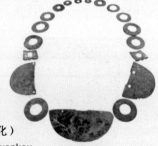

- 玉挂件（大汶口文化）
Jade Pendants (Dawenkou Culture, approx. 4500 years ago)

佛珠：佛教徒念佛经时，用来计算诵经次数的串珠。一般由108颗珠子组成一串，故又叫"百八丸"，少数珠上刻有纹饰，并配有流苏。

Buddha Beads: It's used by Buddhists to calculate the times of chanting a Sutra, were often strung with 108 beads, and therefore also called Hundred-eight-pellet. A few beads were carved with patterns and decorated with fringes.

- 玉佛珠
  Jade Buddha Beads

玉锁：锁形片状的挂饰，多由儿童佩戴。表面刻有文字，通常是"长命百岁""永葆青春""玉堂富贵"等吉祥语。

Jade Lock: It refers to a flat lock-shaped pendant, mainly worn by children. Its surface was engraved with words, which were usually Longevity, Youth Forever, Prosperity or other auspicious phrases.

- 玉锁形佩（现代）
  Jade Lock Shaped Pendant (Modern Times)

扳指：最初是戴在大拇指上便于射箭时勾弦的玉筒状物，后来转化为装饰品，仍戴于大拇指上。

Thumb Ring (*Banzhi*): It refers to a tube shaped jade item, originally worn on an archer's thumb for pulling the string of a bow, which evolved into an ornament, and was still worn on one's thumb.

- 翡翠扳指（清）
  Jadeite Thumb Ring (*Banzhi*)(Qing Dynasty, 1616-1911)

实用玉器

实用玉器包括用于生产的玉工具、盛放器物的玉制器皿，以及各种文房用具等。实用玉器的特点是具有实用价值，但是由于玉料珍贵，所以通常也制作精美，具有较高的艺术价值和收藏价值。

玉工具：用玉琢制而成的生产工具，主要有斧、凿、刀、锛、铲等。有些玉工具也可作兵器使用，同时兼具礼器功能。后来随着时代的发展，青铜冶铸业发达起来，玉质工具逐渐消失。

Practical Jade Articles

Practical jade articles included jade instruments of production, jade vessels for holding objects, all kinds of jade study appliances, and etc. Practical jade articles were characterized by their practical values. Nevertheless, since jade materials were precious, they were often delicately crafted and therefore had high artistic and collection values as well.

Jade Instruments: The tools are made out of jade, mainly including axes, chisels, knives, adzes, shovels and etc. Some jade instruments could also be used as weapons and ritual objects as well. With the advancement of time, bronze casting industry flourished and jade instruments gradually disappeared.

玉工具的种类
Different Kinds of Jade Instruments

玉斧：一般较厚重，刃直，是玉钺、玉戚等其他玉器产生的原始器形。

Jade Axe: It's thick and heavy, with a straight blade, which is the original model for jade *Yue*, jade *Qi* and other jade articles.

• 玉斧（商）
Jade Axe (Shang Dynasty, 1600 B.C.-1046 B.C.)

玉锛：是一种与斧形似的工具，较小，单面刃，通常用来切割或凿削。

Jade Adze (*Ben*): It refers to an axe-like instrument, relatively small, which had a one-sided blade, usually used to cut, scrape or bore.

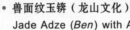

- 兽面纹玉锛（龙山文化）
  Jade Adze (*Ben*) with Animal-Faced Pattern (Longshan Culture, 2500 B.C.-1900 B.C.)

玉铲：外形接近于玉钺，有的像玉圭，通常用来削平或挖取东西。

Jade Shovel: Resembling jade *Yue* or jade *Gui* in appearance, it was usually used to scrape or dig.

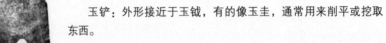

- 玉铲（龙山文化）
  Jade Shovel (Longshan Culture, 2500 B.C.-1900 B.C.)

玉凿：一种窄长体，或平刃或弧刃的工具，通常用来挖槽或穿孔。

Jade Chisel: It refers to a kind of narrow and long instrument with either a level blade or an arc-shaped blade, usually used to dig or bore.

- 玉凿（商）
  Jade Chisel (Shang Dynasty, 1600 B.C.-1046 B.C.)

玉纺轮：一种四周为圆形，中间有穿孔，壁比较厚的玉质工具，用于纺织。

Jade Spindle Wheel: It refers to a perforated, thick disc of jade, used for spinning.

• 玉纺轮（良渚文化）
Jade Spindle Wheel (Liangzhu Culture, 4000-5000 years ago)

玉制器皿：玉制器皿数量大、种类多，常见的有玉炉、玉熏、玉樽、玉瓶、玉碗、玉壶、玉杯、玉盆、玉盒、玉盏等。玉制器皿制作要求用料质色要均一，形状要规矩，所以制作工艺难度较大。

Jade Vessels: large in quantity, various in kind, commonly included jade censers, jade *Xun* (a kind of incense burner to fume away ghosts and evil spirits), jade *Zun* (wine vessel), jade vases, jade bowls, jade pots, jade cups, jade basins, jade cases, jade *Zhan* (small cup) and etc. Jade vessels required uniformity in material, texture and color, and should be regularly shaped, so they were hard to be crafted.

玉制器皿的种类
Different Kinds of Jade Vessels

玉香炉：用玉雕成的用于焚香的器皿，也可以用来熏衣服或是用于室内陈设。一般没有盖，形式多样，有鼎式炉、簋式炉、鬲式炉、亭式炉、冲耳高足炉等。

Jade Incense Burner: It refers to a vessel made of jade used to burn incense, fumigate clothes, or furnish a room. Often uncovered, jade censers were various in shape, including *Ding*-shaped (a cauldron) censer, *Gui*-shaped (a round-mouthed food vessel with two or four loop handles) censer, *Ge*-shaped (a round-bellied food container with three hollow legs) censer, pavilion-shaped censer, ear-like handled (*Chong'er*) high-foot censer and etc.

- 翠夔龙纹双耳亭式炉（清）
  Jadeite Double-handled Pavilion-shaped Censer with *Kui*-dragon (Mythological Animal) Pattern (Qing Dynasty, 1616-1911)

玉瓶：用玉雕成的口小、腹大、体修长的器皿，通常用来盛酒、盛水、盛物等。玉瓶形制较多，常见的器形有方瓶、贯耳瓶、观音瓶、鸡腿瓶、八棱瓶、圆肚瓶、葫芦瓶、仿瓷瓶等。

Jade Vase: It refers to a small-mouthed, big-bellied, slim vessel made out of jade, usually used to hold wine, water, and other objects. Various in shape, jade vases commonly include square vases, vases with pierced handles, avalokitesvara (*Guanyin*) vases, chicken-leg shaped vases, octagonal vases, round belly vases, gourd-shaped vases, vases modeled on porcelain vases and etc.

- 花卉纹玉瓶（清）
  Jade Vase with Floral Pattern (Qing Dynasty, 1616-1911)

- 四耳活环方盖玉瓶（清）
  Jade Four-eared-and-ringed Vase with Square-shaped Cover (Qing Dynasty, 1616-1911)

玉杯：用于饮酒的玉器皿，形制多样。如杯体有圆、扁圆、斗方、多棱、花朵等形状，杯柄有錾形、花形、螭形等，也有的无柄。

Jade Cup: It refers to a jade vessel used to drink wine, various in shape. The cup body could be round, oblate, square (*Doufang*), multi-ridged, and flowery shaped. Some cup's handles were shaped like *Pan* (handle), flowers or *Chi*-dragon, while some cups were handleless.

- 龙形柄桃式玉杯（明）
  Jade Peach-shaped Cup with Dragon-shaped Handle (Ming Dynasty, 1368-1644)

- 金盖托白玉杯（清）
  White Jade Cup with Gold Cover and Saucer (Qing Dynasty, 1616-1911)

玉壶：用于盛酒或盛水的玉器皿，形制多样。有的仿古代青铜器中鼎、豆、鉴、爵、角等器形，有的仿瓷器中的葫芦瓶、双筒瓶等器形，还有的在壶体两侧增加装饰，这一类大多作为陈设之用。

Jade Ewer: It refers to a jade vessel used to hold wine or water, various in shape. Some were modeled on ancient bronze cauldrons (*Ding*), compote (*Dou*), vat (*Jian*), Jue (an ancient wine vessel with three legs and a loop handle), *Jue* (another ancient wine vessel similar to *Jue*, yet without spout), and other vessels, while some were modeled on porcelain gourd-shaped vases, double-tube vases, and etc. Some pots were decorated at both sides, which were mainly used as furnishings.

- 玉壶（明）

  Jade Ewer (Ming Dynasty, 1368-1644)

- 白玉瓜棱扁壶（清）

  White Jade Ewer with Melon-shaped Design (Qing Dynasty, 1616-1911)

玉花插：用于插花的玉器皿，多为筒状，器身常雕刻花纹，以树桩形、花朵形、荷叶形等较为常见。

Jade Flower Receptacle (*Huacha*): It refers to a jade vessel used to arrange flowers, mainly tube-shaped, often carved with patterns that were commonly stump, flower, and lotus-leaf shaped.

- 玉兰花形玉花插（清）

  Jade Magnolia-shaped Flower Receptacle (*Huacha*)(Qing Dynasty, 1616-1911)

玉熏：用来燃香料熏房间或衣服的玉器皿，有盒式、筒式、炉式三种基本造型。

Jade Aroma Diffuser (*Xun*): refers to a jade vessel used to burn spices to scent a room or clothes, which was basically box-shaped, tube-shaped or censer-shaped.

- 白玉筒式香熏（清）
  White Jade Tube-shaped Aroma Diffuser (*Xun*)(Qing Dynasty, 1616-1911)

玉盒：用于盛放东西的玉器皿，一般较小，常见的有方盒、圆盒、桃式盒、荔枝盒、银锭盒、蔗段盒、蒸饼式盒等。

Jade Case: It refers to a jade vessel used to contain items, often small in size. Commonly there were square cases, round cases, peach-shaped cases, lychee-shaped cases, sycee-shaped cases, sugarcane-cut shaped cases, pancake shaped cases and etc.

- 鹌鹑形玉盒（清）
  Jade Quail-shaped Case (Qing Dynasty, 1616-1911)

玉盘：用玉雕刻而成的盘状器皿。商代时就已有玉盘出现，腹较深，应为盛水之用。明清时期的玉盘腹较浅，应为餐具。

Jade Plate: It refers to a plate-shaped vessel made out of jade. As early as the Shang Dynasty, deep-bellied jade plates appeared, which might be used to hold water. During Ming and Qing dynasties, jade plates were shallow-bellied, which might be used as dishes.

- 青玉盘（清）
  Gray Jade Plate (Qing Dynasty, 1616-1911)

玉碗：用来盛放食物的玉器皿。玉碗的制作工艺虽不复杂，但非常费料、费时，所以到明代以后才大量出现。玉碗均为圆形，有敞口、直口两种，碗外刻有多种纹样。

Jade Bowl: It refers to a jade vessel used to contain food. Although jade bowls were not difficult to be crafted, they appeared in large amounts only after the Ming Dynasty (1368-1644), since it had been material-consuming and time-consuming in making. Jade bowls were all round-shaped, either open-mouthed or upright mouthed, carved with various patterns on the surface.

• 青玉刻花碗（明）
Gray Jade Bowl with Incised Floral Pattern (Ming Dynasty, 1368-1644)

玉文房用具：玉文房用具包括砚、笔架、笔筒、笔洗、墨床、印泥盒、臂搁、水盂等。但其中有些不具备实用功能，如玉砚就缺少发墨功能，只能作为陈设品，以供玩赏。

Jade Study Tools: It included ink stone, brush rack, brush pot, brush washer, ink rest, ink pad box, arm rest, water pot (*Shuiyu*) and etc. However, some of them didn't have practical applications. For example, jade ink stone couldn't generate ink, and therefore could only be appreciated as a decorative object.

玉文房用具的种类
Different Kinds of Jade Study Appliances

玉笔筒：用玉雕成的用来盛放毛笔的筒状物。器形以圆筒形居多，笔筒外壁都有浮雕的山水人物，做工非常精细。

Jade Writing Brush Pot: It refers to a pot carved out of jade, used to hold writing brushes, mainly cylindrically shaped, with landscapes and figures in relief on the exterior, which were highly delicately crafted.

- 碧玉九老图笔筒（清）
  Green Jade Brush Pot with the Design of Nine Elders(Qing Dynasty, 1616-1911)

玉书镇：用于压纸张的玉器物，一可防止纸被风吹走，二可把纸抻平，便于写字作画。造型变化虽多样，但都有底部平整、重心低、有一定的重量、表面光洁四个基本特点。

Jade Paperweight: It refers to a jade article used to hold down sheets of paper. First, it could prevent a stack of paper from being blown away. Second, it facilitated stretching the paper for writing or drawing. Although various in shape, jade paperweights had four basic characteristics in common, specifically, a flat bottom, a low center of gravity, relatively heavy weight, and a bright and clean surface.

- 青玉"三阳开泰"镇纸（清）
  Gray Jade Paperweight with the design of *Sanyang Kaitai* (Auspicious Beginning of a New Year)(the Qing Dynasty, 1616-1911)

玉笔架：临时搁放毛笔的玉器物。古代中国人在用毛笔书写时，需要暂时停顿下来，由于笔头沾有墨，毛笔又是圆筒状，不容易搁放，所以笔架是文房中不可缺少的用具之一。

Jade Writing Brush Rack: It refers to a jade article where writing brushes were temporarily placed. When ancient Chinese people stopped for a while during their writing, the cylindrically shaped writing brush with an ink-wetted brush tip could not be easily laid. Therefore, brush rack was one of the indispensible instruments in a study.

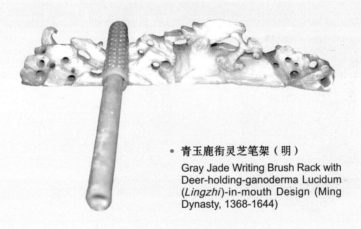

• 青玉鹿衔灵芝笔架（明）
Gray Jade Writing Brush Rack with Deer-holding-ganoderma Lucidum (*Lingzhi*)-in-mouth Design (Ming Dynasty, 1368-1644)

玉砚：用玉雕琢成的砚台，由于缺少发墨功能，所以往往只是一种陈设品。玉砚有凤字形砚、荷叶随形砚、卧鹅式砚、凤背砚等多种形式。

Jade Inkstone: It refers to an ink stone made out of jade, and was often used as a decorative object, since it could not generate any ink. There were jade ink slabs respectively crafted in the shape of the Chinese character "凤", a lotus-leaf, a lying goose, the back of a phoenix, and etc.

• 青玉鹅式砚（清）
Gray Jade Goose-shaped Inkstone (Qing Dynasty, 1616-1911)

玉砚滴：是向砚里注水的工具。一般做成青蛙、乌龟、瑞兽的形状，摆设在书桌上，也极具观赏价值。

Jade Ink Dropper (*Yandi*): It refers to a device used to add water to an inkstone, usually crafted into the shape of a frog, a turtle, or an auspicious animal, displayed on a desk, highly pleasing to the eye.

- 青白玉兽形砚滴（宋）
  Greenish-white Jade Animal-shaped Ink Dropper (*Yandi*)(Song Dynasty, 960-1279)

玉臂搁：用来搁手臂的一种玉文房用具。古代中国人写毛笔字的顺序是从右往左，从上往下，在换行的时候手腕正好在刚写好的字上，容易把字蹭花、把衣服弄脏，臂搁的制作与使用解决了这一问题。

Jade Arm Rest: It refers to a jade study instrument where one's arm was placed. Ancient Chinese people used to write on a sheet of paper, from right to left, top to down. One would easily mess up the characters and smear ink on his clothes when writing on a new line, with the wrist being just over the newly finished characters. This problem was solved by the application of arm rests.

- 翡翠臂搁（清）
  Jadeite Arm Rest (Qing Dynasty, 1616-1911)

玉墨床：在研磨墨停顿下来时用来放墨锭的小玉件。研磨停顿时，墨锭一头沾有墨汁，若随意放很容易弄脏别处，墨床就用来临时搁置墨锭。常见的墨床形状有几形、多宝槅形、盒形等。

Jade Ink Rest: It refers to a small jade article used to hold the ink stick when a person stopped grinding ink for a while. Since the ink stick was ink-wetted where it touched the ink stone, it would easily stain other things if left about. Accordingly, an ink rest served as a utensil where the stick was temporarily placed. An ink rest was usually crafted in the shape of a small table, a curio cabinet, or a case.

- 青玉墨床（清）
  Gray Jade Ink Rest (Qing Dynasty, 1616-1911)

玉印泥盒：用来存放印泥的小玉盒，有盖，盖与器身之间有子母扣。常见的器形有圆形、椭圆形、方形倭角形、竹节形等。

Jade Inkpad Box: It refers to a covered small jade box used to contain an inkpad. Between the cover and the box was a buckle. Usually jade inkpad box was round shaped, oval shaped, square shaped with chamfered corners, or bamboo-knot shaped.

- 白玉雕螭方印盒（明）
  White Jade Inkpad Box with *Chi*-dragon Design (Ming Dynasty, 1368-1644)

玉笔洗：用于清洗毛笔的玉制容器。古代中国人用毛笔蘸墨写字，因墨中含有胶，墨干后会把毛笔的毛黏住，再用水泡开时，会损伤毛笔，所以一般使用完毛笔后，需在笔洗中把毛笔洗净。玉笔洗有荷叶式、贝叶式、葵瓣式、蕉叶式、瓜式、葫芦式、凤式等形状。

Jade Writing Brush Washer: It refers to a jade vessel used to wash writing brushes. Ancient Chinese people wrote with writing brushes by dipping them in the ink. Since the ink contained glue in it, the hairs of writing brushes would be glued together when the ink was dried. When the hairs were soaked in water to be unglued, it would cause some damage to the writing brushes. Therefore, writing brushes need be cleansed in a brush washer after use. A jade brush washer could be lotus-leaf shaped, pattra-leaf shaped, sunflower-petal shaped, banana-leaf shaped, melon shaped, gourd shaped, or phoenix shaped.

• 青玉秋蝉桐叶笔洗（清）
Gray Jade Writing Brush Washer with Autumn Cicada and Candlenut Leaf Design (Qing Dynasty, 1616-1911)

玉水盂：盛水的小型玉容器，磨墨时可用小匙从水盂中舀出水来。一般配有铜或珊瑚制成的匙。器形有方形、圆形、海棠形、如意形、桃形、荷叶形、葫芦形等。

Jade Water Pot (*Shuiyu*): It refers to a small jade vessel that contained water. One could use a scoop made of copper or coral, to ladle water out when grinding ink. Usually, a jade water pot could be square shaped, circle shaped, Chinese small apple shaped, *Ruyi* shaped, peach shaped, lotus-leaf shaped, or gourd shaped.

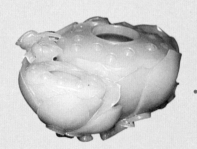

• 莲式玉水盂（清）
Lotus Shaped Jade Water Pot (*Shuiyu*) (Qing Dynasty, 1616-1911)

丧葬用玉

古代生产力低下，人们迷信鬼神，笃信宗教，认为人死了之后，灵魂不会死亡。远古先民认为玉能够保护死者的灵魂，因此产生了以玉敛葬的习俗。在新石器时代各文化遗址中均有随葬玉器出土，多是死者生前的用品或特别喜爱的玉器，但这些还算不上真正的葬玉。葬玉是指专门用来保护死者尸体的玉器，盛行于战国至汉代，主要有玉衣、玉握、玉塞、玉玲等，几乎覆盖了尸体的每一处。

Jade Funerary Articles

In ancient times, due to the low productivity, people were superstitious about ghosts and gods and had strong religious belief, holding that human beings' souls would outlive their bodies. Ancient ancestors believed that jade could protect the souls of dead people, and therefore developed the custom of coffining jade with corpses. From all kinds of the Neolithic Age cultural relics have excavated jade articles buried with the corpses, which were mainly used or adored by the deceased before their death, but these were not jade funerary articles in the real sense. Jade funerary articles refer to specially designed jade articles used to protect a corpse, prevalent from the Warring States Period to the Han Dynasty, which mainly included jade clothes, jade *Wo*, jade *Sai* (plug), jade *Han* and etc., almost covering every single place of a corpse.

丧葬用玉的种类
Different Kinds of Jade Funerary Articles

玉衣：汉代皇帝和贵族专为罩尸用的玉制葬服，比真人体形略大，按部位分为头罩、上身、袖子、手套、裤筒和鞋子六部分。东汉时有严格的玉衣制度，皇帝用金缕玉衣，诸侯王、列侯、公主用银缕玉衣，长公主用铜缕玉衣。

Jade Suit: It refers to a burial suit made of jade used to cover the deceased emperors and aristocrats during the Han Dynasty (206 B.C.-220 A.D.), larger than a real human body, which was divided into six parts, specifically, a helmet, the upper part, sleeves, gloves, trousers and shoes. During the Eastern Han Dynasty (25-220), there was a rigid system for jade clothes. Jade clothes sewn with golden thread were used for emperors; those with silver thread were used for dukes, noblemen and princesses, and those with bronze thread were used for princesses royal.

- 窦绾金缕玉衣（汉）
  Dou Wan's Jade Burial Suit with Golden Thread (Han Dynasty, 206 B.C.-220 A.D.)

玉塞：又叫"九窍塞"，即填塞死者身上九窍孔的九件玉器，分别是耳塞二件、眼塞二件、鼻塞二件、口塞、肛塞和生殖器塞各一件。

Jade *Sai*: It's also called nine orifices plugs (*Jiuqiao Sai*), refer to nine jade articles used to plug the nine orifices of a corpse, which were respectively a pair of earplugs, a pair of eye-plugs, a pair of nasal plugs, a mouth plug, an anal plug, and a genital plug.

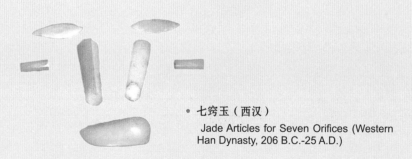

- 七窍玉（西汉）
  Jade Articles for Seven Orifices (Western Han Dynasty, 206 B.C.-25 A.D.)

玉握：指握在死者手中的玉器。

Jade *Wo*: It refers to jade articles held in the deceased's hands.

• 白玉握（东汉）
White Jade *Wo* (Eastern Han Dynasty, 25-220)

玉琀：是死者含在口中的葬玉，多为蝉形，故又称"琀蝉"，形制也较为简单，具有"汉八刀"风格。

Jade *Han*: It refers to a funerary jade article put into the mouth of the deceased, mainly cicada shaped, therefore also called jade cicada, whose shape was simple, displaying a style of *Hanbadao* (referring to the craftsmanship of shaping an object with several cuts).

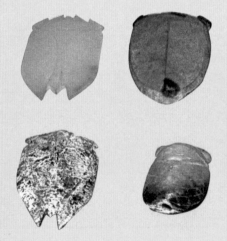

• 玉琀蝉（西汉）
Jade Cicada Shaped Han (Western Han Dynasty, 206 B.C.-25 A.D.)

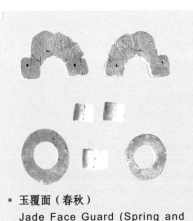

玉覆面：是放于死者面部的玉片组合。通常由十几片玉片打孔后穿缀组成，按照五官的形状分别琢磨成眉、目、鼻、口、耳的形状，然后用丝线连缀起来，形成与面部相吻合的玉面罩。

Jade Face Guard: It refers to a set of jade tablets covering the deceased's face, usually stringed with a dozen perforated jade tablets or so, respectively crafted into the shapes of eyebrows, eyes, nose, mouth, and ears. These tablets were connected with silk thread, forming a jade face guard that matched the face.

- 玉覆面（春秋）
Jade Face Guard (Spring and Autumn Period, 770 B.C.-476 B.C.)

陈设用玉

陈设用玉的主要用途是陈设玩赏。总体来说，陈设用玉器形较大，造型美观，常见的有玉山子、玉座屏、玉人物摆件、玉兽摆件、玉花鸟摆件、玉如意等玉摆件。

Jade Furnishing Articles

Jade furnishing articles were mainly used to be displayed and appreciated. Generally, jade furnishing articles were large in size, beautifully shaped, which commonly included jade *Shanzi* (jade display with landscape design), jade screens, jade ornaments respectively with figures, beasts, and flowers and birds design, jade *Ruyi* (a curved decorative object symbolizing good fortune) and etc.

陈设用玉的种类
Different Kinds of Jade Display Articles

玉山子：指将玉料加工成山林景观形状的玉雕，因玉料大小而巨微不一，大的可达数吨重、一米多高，小的则只有十几厘米大小。

Jade *Shanzi*: It refers to a jade carving with the appearance of a landscape, whose size, based on its jade material's size, could be as large as several tons in weight and more than one meters in height, or as small as ten centimeters or so in height.

- 碧玉山聚珍图（现代）
  Green Jade Mountain with a Collection of Rarities (Modern Times)

玉座屏：指用大块薄片玉板为屏心，雕刻纹样之后，插于硬木屏座上而形成的玉摆件。玉屏心有方形、长方形、圆形等形状，雕刻题材有花鸟、名树、著名风景、历史典故、吉祥图案、百寿字、名诗等。

Jade Screen: It refers to a jade ornament, which was a large thin sheet of jade carved with patterns, installed on a screen stand made of hardwood. A jade screen could be square, rectangular and circular shaped with the carving motifs ranging from flowers and birds, renowned trees, famous landscapes, historical allusions, auspicious patterns, hundred writing forms of *Shou* (Chinese character, meaning longevity), to well-known poems.

- 青玉宴饮图玉座屏（清）
  Gray Jade Screen with Banqueting Design (Qing Dynasty, 1616-1911)

玉人物摆件：是以人物造型为主的玉摆件，题材有佛像、神话人物、仕女、老人、小孩、历史人物等。人物造型注重身体比例、姿势、脸部的刻画以及人物的神态等内容，以圆雕为主。

Jade Displays with Figures Design: It refers to jade ornaments modeled on figures, whose motifs included figures of Buddha, legendary figures, beautiful women, the aged, children, historical figures and etc. These statuettes were delicately crafted, with emphasis on body proportions, postures, faces and countenances, mainly in full relief.

- 玉仙姑像（清）
Jade Statuette of a Celestial Figure (*Xiangu*)(Qing Dynasty, 1616-1911)

玉兽摆件：是以瑞兽和写实兽造型为主的玉摆件，也以圆雕为主。瑞兽包括龙、麒麟、貔貅等神话传说中的神兽，写实兽指牛、鹿、象、马等自然界中存在的动物。

Jade Displays with Beasts Design: It refers to jade ornaments modeled on auspicious beasts and realistic animals, mainly in full relief as well. Auspicious beasts included dragon, Kylin, *Pixiu* and other mythological beasts, while realistic animals refer to oxen, deer, elephants, horses and other authentic animals in the real world.

- 白玉象（清）
White Jade Elephant (Qing Dynasty, 1616-1911)

玉雕花鸟摆件：以花鸟为形雕制而成的陈设玉件，题材广泛，供选用的花卉多样，常选用的有牡丹、月季等，也有寓意吉祥如意的组合花卉，在花卉中常配有鸳鸯、鹤、孔雀等鸟类图案。

Jade Displays with Flowers and Birds Design: It refers to jade ornaments modeled on flowers and birds, various in carving motif. There was a wide range of flowers designs, from commonly used peonies and Chinese roses, to mixed flowers symbolizing auspiciousness. Flowers designs were often joined with mandarin ducks, cranes, peacocks and other birds patterns.

- 碧玉绶带鸟摆件（清）
  Green Jade Display with Paradise Flycatchers Design (Qing Dynasty, 1616-1911)

玉如意：供玩赏之用的吉祥器物，有长柄钩形、灵芝形、云头形等多种形状。玉如意上一般雕有精美的寓意吉祥的纹饰。玉如意也是宫廷必备之物，到清代时成为宫廷重要的礼器之一。每逢宫廷有重要的喜庆之事，如意都会受到重用。

Jade *Ruyi*: It refers to an auspicious article used for appreciation, including long-handled-hook shape, ganoderma lucidum (*Lingzhi*) shape, cloud shape, etc. Jade *Ruyi* was usually carved with exquisite patterns symbolizing auspiciousness, which was a must for the imperial court, becoming one of the court's most important ritual objects during the Qing Dynasty (1616-1911). Once the court held major festivals, *Ruyi* would be put in important positions.

- 青玉雕八宝纹如意（清）
  Gray Jade Ruyi with Eight-treasure Pattern (Qing Dynasty, 1616-1911)

玉器的制作

玉器造型变化多端，但玉器的制作流程是差不多的，即选料→设计→琢磨→抛光。

选料，是玉器制作的第一道工序，通过正确合理地选用玉石原料，确定原料适合做什么产品，达到合理使用玉料的目的。

设计，是根据玉料特点设计造型，突出原料的质地、光泽、颜色、透明度等特质，充分展示玉质美。

琢磨，是利用砣和磨粉等制玉工具，按设计图加工成产品。

抛光，是将玉件的粗糙部位，碾磨平整，并通过应用一些化学粉剂原料作介质，使玉件显露出玉材光洁、温润和晶莹的本质。

Production of Jade Articles

Although various in shape, jade articles are crafted through a similar production process: material selection → design → carving and grinding → polishing.

Material selection is the first procedure for making jade articles. Jade material is correctly and appropriately selected and decided on what kind of article can be suitably made into, for the purpose of effectively using the material.

Design refers to designing a shape according to the characteristics of a particular jade material, which can help to intensify the jade material's texture, luster, color, transparency and other properties and fully display the beauty of the jade's quality.

Carving and grinding, refers to processing jade articles according to the design, by using emery wheels, grinding powder, and other jade-making instruments.

Polishing refers to rubbing off the rough edges of jade articles and manifesting the bright and clean, mild and crystal-clear nature of jade material by applying some chemical powder as media.

• 按照设计图加工玉器
Processed Jade Article According to the Design

> 纹饰之美

玉器上的纹饰题材丰富，有动物、植物、几何、人物、文字和复合型图案等。纹饰是玉器的重要元素，不仅为玉器增加了美感，也赋予了玉器丰富的文化内涵。玉器纹饰中有许多来自人们的信仰、民间传说、动植物的谐音等吉祥图案，是中国传统文化的一种表现形式。

几何纹

几何纹是一种最原始的纹饰，是由点、线、圈等图案组成的装饰题材。它主要包括谷纹、蒲纹、连珠纹、弦纹、涡纹、绳纹、乳钉纹、云纹、云雷纹等。

> The Beauty of Pattern

The patterns on jade articles are various in motifs, ranging from animals, plants, geometry, figures, words, to complex designs. Pattern is an important element of jade articles, not only increasing the beauty of jade articles, but also endowing jade articles with rich cultural connotations. The patterns on jade articles mostly originate from people's beliefs, legends, animals and plants that bear homonymic names (with propitious phrases) and other auspicious patterns, which serve as a form of expression of traditional Chinese culture.

Geometric Pattern

Geometric pattern is the most primitive pattern, referring to a decorative motif consisted of dots, lines, circles and other geometric figures, mainly including millet pattern, cattail pattern, multi-cycle pattern, string pattern, swirl pattern, cord pattern, studs pattern, cloud pattern, thunder cloud pattern and etc.

几何纹的种类
Different Kinds of Patterns

谷纹：是由分布密集的圆形凸起和呈旋涡状的圆点组成，看起来似蝌蚪。谷纹象征着万物复苏、生机勃勃的景象和人们对春天的盼望。

Millet Pattern: It consists of densely distributed circular bulges and vortex-like dots, appears like tadpoles, which symbolizes the recovery of the natural world, vibrant prospects, and people's longing for spring.

- 谷纹玉璧（汉）
Jade *Bi* with Millet Pattern (Han Dynasty, 206 B.C.-220 A.D.)

蒲纹：是一种成排而列的六角形格子形纹样，看起来如同编织的蒲席。古代中国人习惯"席地而坐"，即坐在用蒲草编织的席子上。蒲纹表达出人们对于安居乐业的向往和祈求。

Cattail Pattern: It referring to a lined hexagonal lattice-shaped pattern, appears like a piece of woven cattail mat. Ancient Chinese people used to sit on the ground, more specifically, on a cattail mat. Accordingly, cattail pattern conveyed people's yearning and prayer for a harmonious household.

- 蒲纹玉璧（汉）
Jade *Bi* with Cattail Pattern (Han Dynasty, 206 B.C.-220 A.D.)

连珠纹：是用小圆圈作横式排列而产生的纹样，一般作为主纹的分栏线或作为边饰衬托主纹。

Multi-circle Pattern: It is consisted of small circles horizontally arrayed, usually used as a column-dividing line for the main patterns or border ornaments for intensifying the main patterns.

- 云鹤连珠纹玉带板（明）
Jade Belt Tablet with Crane and Multi-circle Pattern (Ming Dynasty, 1368-1644)

弦纹：一根根凸起的直线条，是玉器上常见的辅助纹，也有的玉器纹饰只有弦纹。

String Pattern: It consisted of rows of bulging straight lines, is a common auxiliary pattern on jade articles. Some jade articles are only decorated with string pattern.

- 弦纹玉璧（商）

Jade *Bi* with String Pattern (Shang Dynasty, 1600 B.C.-1046 B.C.)

涡纹：一种近似水涡旋转的几何纹样，外围是一个较大的圆环，自外向内旋出多组旋线，涡纹常作为辅助纹出现，一般用于小件玉器上。

Swirl Pattern: It refers to a geometric pattern appearing like a whirling vortex. There is a relatively big circle on the periphery, from where multi-vortex lines were drawn toward the center. Whorl pattern usually appears as an auxiliary pattern, often used on small jade articles.

- 涡纹玉饰片（战国）

Jade Decorative Tablet with Swirl Pattern (Warring States Period, 475 B.C.-221 B.C.)

绳纹：是一种由两条以上的波线纹交织扭结成绳索状的纹样，有粗有细，多饰于玉器分界处，以及动物形器物的眉、尾部。

Cord Pattern: It made by twisting more than two wave-like cords into a rope-like pattern, can be various in thickness, usually used to decorate the border area of a jade article, and an animal-shaped article's eyebrows and tail areas.

- 绳纹青玉镯（明）

Gray Jade Bracelet with Cord Pattern (Ming Dynasty, 1368-1644)

乳钉纹：是一种形状为凸起的乳突状圆钉的纹样，圆钉排列形式不一，或纵横排列，或不规则排列，或琢在蒲纹交叉线所构成的空格里。乳钉纹最早出现在祭祀女性先人的祭器之上，表示对母亲的敬仰和怀念，也有祈求子孙满堂、人丁兴旺的寓意。

Studs Pattern: It appears like mastoid round studs. The round nails can be variously arrayed, in the horizontal and vertical way, or an irregular way. They can also be made within the blank spaces, formed by the crossing lines of cattail pattern. Studs pattern originally appeared on the ritual objects used to offer sacrifice to female ancestors, which conveyed one's respect and pining for mother, and symbolized one's prayer for a thriving household.

- 青玉乳钉纹龙耳簋（明）
  Gray Jade *Gui* (Round and Eared Food Container) with Studs Pattern and Dragon-shaped Handles (Ming Dynasty, 1368-1644)

云纹：是古人用以刻画天上之云的纹饰，根据其形态特征可分为单岐云（一朵云）、双岐云、三岐云、勾云纹、云雷纹和云水纹等。古人以农耕为主业，农业生产全靠雨露滋润，无云便无雨，云代表雨，云纹有敬天、求雨的寓意。

Cloud Pattern: It crafted by ancient people to depict the clouds in the sky, can be divided into single *Qiyun* (one cloud), double *Qiyun*, three *Qiyun*, curving cloud, thunder cloud and water cloud patterns. Ancient people lived on agriculture, which totally depended on the rain for irrigation. No clouds, no rain. Clouds herald the rain. Therefore cloud pattern symbolized people's respect for the heaven and prayer for the rain.

- 云纹螭耳玉匜（明）
  Jade *Yi* (Ancient Ritual Object Used with Plate for Hand-washing, Having a Flat Oval Body, with a Spout and a Handle, Four-legged) with Cloud Pattern and *Chi*-dragon-shaped Handle (Ming Dynasty, 1368-1644)

动物纹

动物纹是玉器上常见的纹饰，一般是以具体的动物形象为题材，加以图案化处理的纹饰。常见的动物纹有兽面纹、夔龙纹、螭纹、蟠螭纹、虺纹、麒麟纹、龙纹、凤纹、虎纹、鹿纹、鱼纹、龟纹、蛇纹等。

Animal Pattern

Animal pattern is a common decoration on jade, and usually has a specific animal prototype processed into abstract patterns. Common animal patterns are animal-faced pattern, *Kui*-dragon pattern, *Chi*-dragon pattern, *Pan-Chi* pattern, *Hui* (snake) pattern, Kylin pattern, dragon pattern, phoenix pattern, tiger pattern, deer pattern, fish pattern, tortoise pattern and snake pattern, etc.

动物纹的种类
Types of Animal Pattern

兽面纹：又称"饕餮纹"。饕餮是古代传说中的一种贪食的恶兽。兽面纹以鼻梁为中轴线，两侧五官对称排列。纹饰的最上面有角，角下有突出的兽目（有些兽目下还有眉），兽目两侧有耳，鼻翼两侧是张开的大嘴，有的嘴里还有几颗尖牙。在兽面的两边有曲折向上的躯体，躯体下有腿和爪。也有的兽面纹没有躯体和利爪。

Animal-faced pattern: It is also known as *Taotie* pattern, which is a ferocious and gluttonous animal in the ancient legend. In this design, the facial features of the beast, with nose to the central axis, are symmetrically arranged on both sides. At the top of the decoration, there is the horn; under the horn are the prominent eyes of the beast (some animals have eyebrows under their eyes); on both sides of eyes are ears and under the open nostrils there is the open mouth, and some animals have fangs in their mouths. On both sides of animal face there is the winding up of a body, and under the body there are legs and claws, nevertheless there are animal mask designs without body or claws.

- 兽面纹玉饰（战国）
Jade Ornament with Animal-faced Design (Warring States Period, 471 B.C.-225 B.C.)

螭纹：螭是一种传说中的动物，其身体和腿似龙，细长，方嘴，肥臀，面部似兽，无角，卷尾。

Chi-dragon Pattern: It is a hornless legendary animal with its body and legs resembling the Chinese dragon, with a long and slender body, a square month, a fat buttock, a coiled tail and a beast-like face.

- 青白玉螭纹鸡心佩（魏晋南北朝）
  Heart-shaped Greenish-white Jade Pendant with *Chi*-dragon Pattern (Wei, Jin, Southern and Northern Dynasties, 220-581)

蟠螭纹：蟠螭指没有角的龙。蟠螭纹类似一种盘曲的螭纹，呈半圆形或近圆形盘曲。

Pan-Chi Design: *Pan-Chi* is similar to a kind of coiled *Chi*-dragon pattern, presenting a circular or semicircular pattern.

- 青玉蟠螭纹佩（清）
  Gray Jade Pendant with *Pan-Chi* Design (Qing Dynasty, 1616-1911)

夔龙纹：夔龙是传说中的神异动物，据文献记载夔是一只脚的龙。夔龙纹的形象多为张口、尾部卷曲。

Kui-dragon Pattern: It is a legendary miraculous animal. According to the ancient literature, *Kui* is a single-foot dragon. This pattern often features it as mouth opened, tail coiled.

- 白玉镂雕夔龙纹出廓佩（西汉）
  Openwork White Jade *Chukuo* (a jade sculpting, literally Means Cross the Outline, Indicating an Advanced Style with Bulging Openwork Dragon, Phoenix or Tiger Design Carved along the Rim of the Jade Ware) Pendant with the *Kui*-dragon Design (Western Han Dynasty, 206 B.C.-25 A.D.)

虺纹：虺是古代传说中一种有剧毒的小蛇，头部较宽大，双眼突出，躯体盘曲，带有鳞节装饰，尾部较细并向上卷。很多盘曲的虺缠绕在一起，便是蟠虺纹。

Hui Pattern: *Hui* is a kind of small poisonous snake in ancient legend. It has a large wide head, protruding eyes, a twisting body, and is decorated with scales, with its tail winding up. Many *Hui*s entwining together forms the *Pan-Hui* pattern.

- 蟠虺纹玉璜（春秋）
  Jade *Huang* (Semi-annular Jade Pendant) with *Pan-Hui* Design (Spring and Autumn Period, 770 B.C.-476 B.C.)

麒麟纹：麒麟是中国神话传说中的瑞兽，性情温良，形象为鹿身，头上有一只角，角上有肉，牛尾巴，马蹄，鱼鳞皮。

Kylin Pattern: It is an auspicious animal in Chinese legends, with a gentle temperament, a deer-like body, an ox-like tail, a horse-like hoof, fish-like scales and a horn on its head.

- 麒麟纹白玉佩（清）
  White Jade Pendant with Kylin Design (Qing Dynasty, 1616-1911)

龙纹：龙在中国神话传说中是一种神异的动物，体长威猛，能腾云驾雾，会兴风降雨。龙纹的形象一般为蛇身，有的有足，有的无足，素面，背部饰有鳞纹。龙是中华民族的图腾，在古代还是帝王权力和地位的象征，故被视为最珍贵的纹样。

Dragon Pattern: Dragon is a miraculous animal in Chinese legends, having a large and mighty body, able to mount the clouds and ride the mist, with the power of unleashing wind and rain. The dragon pattern usually has a snake-like body; some have feet while others do not; with plain face and a scale decorated back. Dragon is the totem of the Chinese nation and the symbol of ancient imperial power and status, thus the dragon pattern is considered the most valuable decoration.

- 龙形玉环（汉）
  Jade Ring with Dragon Design
  (Han Dynasty, 206 B.C.-220 A.D.)

虎纹：虎在中国人的传统观念中是正义、勇猛、威严的象征，被认为是百兽之王。玉器上的虎纹根据现实生活中虎的形态分为两种：一种为虎头纹样，额顶有一对环状竖耳；另一种为虎爬行、目侧视的纹样。

Tiger Pattern: Tiger is a symbol of justice, bravery, and majesty in the traditional Chinese concept, and is considered as the King of Beasts. According to the tiger's shape in real life, the tiger patterns on jade wares are divided into two types: one features a tiger's head, with a pair of standing ears on top of the forehead, the other features a crawling tiger, with its eyes side-looking.

- 青黄玉镂雕虎纹璜（春秋）
  Openwork Greenish-yellow Jade Huang with Tiger Pattern (Spring and Autumn Period, 770 B.C.-476 B.C.)

凤纹：凤在中国传统文化中被认为是吉祥之鸟，为百鸟之王，体型优美，羽毛亮丽。凤纹是以鸟的形象为基础，综合了其他动物外形而创造出来的图案，鸟身，圆眼，长喙，身躯较长，长尾。

Phoenix Pattern: In traditional Chinese culture, phoenix is considered to be an auspicious bird, and the king of birds, with a beautiful body and bright feathers. The phoenix pattern, based on a bird's image, is a combination of many animals' features. It has a bird's body, round eyes, a long beak, a slender body and a long tail.

- 凤纹玉璧（西周）
  Jade *Bi* with Phoenix Pattern (Western Zhou, 1046 B.C.-771 B.C.)

象纹：象在古代中国是吉祥的象征，兆示天下太平。象多以长鼻向上卷曲、牙齿外露、侧身行走的形象出现。象经常与瓶、万年青或童子等组成"太平有象""万象更新""童子涤象"等题材。

Elephant Pattern: In ancient China, elephant (*Xiang*) is an auspicious symbol of world peace and harmony. It usually appears as walking sideways with its trunk curling up and teeth sticking out. The image of elephant usually appears together with bottle (*Ping*), an evergreen (*Wannianqing*) or a boy (*Tongzi*) to homophonically indicate special themes respectively as *Taiping Youxiang* (peace and harmony), *Wanxiang Gengxin* (everything takes on a new look), and *Tongzi Dixiang* (A boy bathing an elephant).

- 玉象（宋）
  Jade Elephant (Song Dynasty, 960-1279)

鱼纹：鱼纹中常见的是鲤鱼和金鱼。鲤鱼的图形多用来表现"连年有余""鱼龙变""鲤鱼跳龙门"等吉语；金鱼的图形多用来表现"金玉满堂""富贵有余"等吉语。

Fish Pattern: The carp and the goldfish are among the most common fish (*Yu*) patterns. The image of carp is usually used to represent those auspicious words and themes, such as *Liannian Youyu* (have surplus food in successive years), *Yu Long Bian* (fish changes into dragon), and the Carps Jumping over the Dragon Gate, etc. The image of goldfish (Jinyu) is also used to express auspicious meanings such as *Jinyu Mantang* (treasures fulfill the house) and *Fugui Youyu* (having superfluous wealth).

● 白玉鱼形佩（明）
Fish-shaped White Jade Pendent
(Ming Dynasty, 1368-1644)

龟纹：龟的寿命长，在中国是长寿的象征。龟纹的形象为四足趴伏，伸头，拖尾，背上光素或满缀圆斑或涡纹等。

Tortoise Pattern: As tortoise has a long life, it is a symbol of longevity in China. It appears as bending over on its four feet, with head out, tail trailing and a plain back or a back decorated with spots and swirl patterns.

● 白玉龟
White Jade Tortoise

鹿纹：鹿在冬天脱角、在春天孽茸，有报春的物候意义。玉器中的鹿纹多作回头状，有"十鹿九回头"之说。另外，"鹿"与"禄"谐音，以鹿寓"禄"。鹿还有长寿之意，与鹤组合，寓意健康长寿。

Deer Pattern: Deer sheds its buckhorn in the winter, and grows antlers in the spring, thus it is a harbinger of spring. The deer pattern on the jade wares is usually described as turning round its head, with the saying of "nine deers out of ten will turn round their heads". In addition, the Chinese characters deer (*Lu*) and prosperity (*Lu*) are homophones, therefore, deer is used to represent prosperity. Deer also has the connotation of longevity, usually combined with the crane, together symbolizing health and longevity.

• 鹿纹饰（明）
Jade Ornament with Deer Design
(Ming Dynasty, 1368-1644)

羊纹：羊在古代中国寓意吉祥。三只羊在一起的图形，寓意"三阳开泰"，表示冬去春来，阴消阳长，预兆吉庆如意。

Ram Pattern: In ancient China ram is an auspicious animal. The image of three-ram represents *Sanyang Kaitai*, which means auspicious beginning of a new year, indicating that the winter has passed and the spring has arrived, meaning that the negative principle in nature (*Yin*) wanes and the positive principle in nature (*Yang*) waxes, and as a symbol of a happiness and success.

• 白玉"三阳开泰"摆件（清）
White Jade Display with the Theme of *Sanyang Kaitai* (Auspicious Beginning of a New Year)(Qing Dynasty, 1616-1911)

植物纹

植物纹，通常以具体的植物形象为题材，纹饰多为几种不同的植物图形组合而成的具有特殊寓意的图案。常见的植物纹有缠枝纹、折枝纹、三多纹等。

Plant Pattern

A plant pattern, usually based on a specific plant image, appears in the form of several different plant patterns combined together to express a special message. The twisting-branch pattern, the folding-branch pattern, the three-abundance pattern are among the common kinds.

植物纹的种类
Types of Plant Pattern

缠枝纹：是一种由藤蔓、卷草演化而成的纹饰，因它的枝蔓相互缠绕，连绵不断，有生生不息之意。缠枝纹以植物的枝或蔓藤为骨架，有缠枝莲、缠枝牡丹等形式。

Twisting-Branch Pattern: It is based upon vines and curly grass. As the branches and tendrils are twisted and continuous, this pattern has the connotation of the endless circle of life. It uses branches and tendrils of plants as the skeleton, and develops a twisting-branch and lotus pattern or twisting-branch and peony pattern, etc.

- 青玉透雕缠枝花卉纹熏（明）
  Openwork Gray Jade *Xun* (a Kind of Incense Burner) with Twisting-branch and Floral Pattern (Ming Dynasty, 1368-1644)

折枝纹：是指截取某种花草的一枝（棵）或部分作为装饰的纹样，一般有花朵和叶子。

Folding-branch Pattern: It refers to cutting out a certain branch or part of a plant as the decoration pattern. It usually has flowers and leaves.

- 青白玉折枝花形佩（金）
Flower-shaped Greenish-white Jade Pendant with Folding-branch Pattern (Jin Dynasty, 1115-1234)

三多纹：是由佛手、蟠桃、石榴或莲子组合而成的纹样，寓意多福、多寿、多子。

Three-abundance Pattern: It comprises of three clusters of plants: bergamot, peach and pomegranate or lotus seed, respectively implying happiness, longevity and abundant descendants.

- 白玉三多纹佩（清）
White Jade Pendant with Three-abundance Pattern (Qing Dynasty, 1616-1911)

人物纹

人物纹，是指以人物活动或人物形象为内容的纹饰，以狩猎、童子、仕女等题材较为常见。狩猎题材一般为猎手纵马驰骋，弯弓持杖，追逐猎物；童子题材则为童子斗草、乐舞的场景，表现了儿童的天真活泼；仕女题材有乐伎、戏婴、梳妆、游乐等。另外还有飞天、菩萨、罗汉、释迦牟尼佛像等人物形象。

Human Figure Pattern

Human figure pattern refers to the decoration featuring human activities and images. Among the most often seen are hunting, children (*Tongzi*), and maids. In the hunting theme, the hunter rides the horse, holding the bow and the stick, chasing his preys. In the children theme, it is usually a child playing on the grass, or singing and dancing, which represents the innocence and liveliness of children. In the maid theme, we have the lady musicians, women playing together with infants, maid doing their make-up and strolling about, so on and so forth. Besides, there are figures of *Feitian*, Bodhisattva, Buddhist Arhat and Buddha.

- 白玉持荷花童子（明）
  White Jade Display with Boy (*Tongzi*)-holding-the-lotus-flower Design (Ming Dynasty, 1368-1644)

- 玉仕女摆件（清）
  Jade Display with Maid Design (Qing Dynasty, 1616-1911)

其他纹饰

除了几何纹、动物纹、植物纹、人物纹，还有一些具有吉祥意义的文字也常作为纹饰，最常见的有福、禄、寿、喜字。另外，由具有特殊寓意的器物组合成的纹饰也很常见，如八宝纹、暗八仙纹等。

Other Types of Patterns

Besides geometric pattern, animal pattern, plant pattern and human figure pattern, there are some using auspicious Chinese characters as the decoration. The characters *Fu* (bliss), *Lu* (prosperity), *Shou* (longevity), and *Xi* (happy) are among the most often seen. In addition, the patterns combined by several objects with special meanings are also common, such as the Eight-treasure pattern and the pattern of emblems of the Eight Immortals.

其他常见纹饰
Other Common Types of Patterns

寿字纹：寿字纹表达的是祝福长寿的愿望。其表现方式主要有两种：一种是将寿字书写得错落有致犹如花纹；另一种是将寿字作图案化布局，作为装饰画面的组成部分。

Shou (longevity) Pattern: It expresses the wish for longevity, which is mainly divided into two forms: one is to make a patchwork of the characters as a decorative pattern; the other is to arrange the characters in a graphic image and use it as a part of the whole picture.

- 青玉雕填金寿字茶碗（清）
  Gray Jade Tea Cup Carved and Filled in with Golden Chinese Character "*Shou*" (Longevity)(Qing Dynasty, 1616-1911)

八宝纹：八宝指佛教常用的象征吉祥的八件器物，即法螺、法轮、宝伞、白盖、莲花、宝瓶、金鱼和盘长结。

Eight-treasure Pattern: It refers to the eight objects used by Buddhists as auspicious omens, namely the triton, the dharma chakra, the treasure umbrella, the white canopy, the lotus flower, the treasure vase, the goldfish and the Chinese knot.

- 青玉八宝纹如意（清）
  Gray Jade *Ruyi* (Furnishing Articles that Symbolize Auspiciousness) with the Eight-treasure Pattern (Qing Dynasty, 1616-1911)

暗八仙纹：暗八仙纹随着八仙纹的出现而流行，在清代玉器上较为常见，"暗八仙"是指八仙所持的器物，分别是汉钟离的芭蕉扇、吕洞宾的宝剑、张果老的渔鼓、曹国舅的阴阳板、铁拐李的葫芦、韩湘子的长箫、蓝采和的花篮、何仙姑的荷花，寓意富贵长寿、平安吉祥。

Pattern of Emblems of the Eight Immortals: It became popular along with the emergence of the Eight Immortals pattern, and it is usually seen on jade articles of the Qing Dynasty (1616-1911). The emblems of the Eight Immortals refer to Han Zhongli's palm-leaf fan, Lv Dongbin's sword, Zhang Guolao's *Yugu* (a percussion instrument made of bamboo), Cao Guojiu's *Yin-Yang* board, Tieguai Li's gourd, Han Xiangzi's long flute, Lan Caihe's flower basket, He Xiangu's lotus flower. Those objects symbolize prosperity, longevity, safety and good fortune.

- 白玉镂空暗八仙纹圆牌（清）
  Openwork White Jade Round Tablet with the Pattern of Emblems of the Eight Immortals (Qing Dynasty, 1616-1911)

- 和田玉"春江花月夜"摆件（现代）
Khotan Jade Display with the Theme of A Moonlight on the Spring River(Modern Times)

选玉技巧
The Skills to Choose a Jade Article

目前玉器市场出现了收藏对象多样化的趋势，不仅有珍稀古玉，还有许多精美的现代玉器可供选择。玉器的选购是一门很大的学问，不仅要学会辨别玉料的真伪、确定玉器的年代，更要具有独到的审美能力和估值能力。

There is a trend of diversification of collections in the jade market; not only rare antique jades, but also many exquisite modern jade wares are all there for customers to choose. To purchase and collect jade is not an easy thing. One has to tell the authenticity and the production year of the jade, and also he must have a unique aesthetic view and the capability to evaluate.

> 如何选购古玉

　　古玉器内涵丰富，能满足收藏者的心理需求，而且古玉器升值空间大，是投资者的最爱，所以古玉向来是玉器购买者的首选。但是玉器市场鱼龙混杂，因此购买古玉必须要掌握古玉的时代特征，了解仿古玉、古玉的作伪手法等，才能购买到心仪的古玉。

　　玉器是一定工艺条件下的产物，带有特定时代的印记。受到社会风尚、审美标准的影响，不同时代的玉器具有不同的器形、纹样、工艺特征。

　　器形：中国自新石器时代开始出现玉制品，至今已有数千年历史，历代玉器造型可以说是千姿百态。器形是判定玉器年代的重要因素之一。

　　新石器时代玉器以片状或圆形

> How to Buy an Antique Jade Article

Rich in culture, antique jades are better able to meet the needs of collectors, and there is a broad space for the value rise, thus antique jades are the favorite for investors and always the first choice for jade purchasers. However, the jade market was so mixed that, in order to buy them a gratifying antique, collectors must be a master of temporal characteristics of jades and know of fake antique jades and forgery techniques. Jade is a product of certain craft conditions, with a particular impression of an era. Influenced by social trends and aesthetic standards, jade articles of different times have different shapes, decorations, and craft characteristics.

　　Shapes: Jade has a long history of thousands of years; since the Neolithic Period (4500-8500 years ago), jade

古代玉器的收藏

中国自古就有尚玉、藏玉的传统，早在夏商周时期，一些有身份的人，在生前佩玉，死后还把玉器一同陪葬。汉代制作了大量精美的玉器，深受统治阶层的喜爱和珍藏。三国两晋南北朝时期，社会动荡，玉器生产停滞，收藏情况也不繁盛。直到隋唐时期，玉器生产开始复苏，出现了富有生活气息、适合民众使用的玉佩饰、玉器皿，玉器走向世俗化。

宋代玉器的生活化趋势更加明显，加之金石学（中国传统文化中的一类考古学，主要研究对象是古代的铜器和碑石铭刻）的兴起，促进了玉器的鉴赏和收藏。收藏玉器蔚然成风，宫廷制玉、藏玉，民间也进行玉器的交易。

明代是玉器发展的繁荣期，玉器制作和收藏规模空前。从北京定陵出土了大批明代玉器，可以窥见一斑。在城市里涌现了大批出售玉器的古董店和集市，还出现了能以假乱真的仿古玉。

清代玉器进入鼎盛期。爱玉成癖的乾隆皇帝，不惜花费大量人力、财力，派人到全国各地收集玉器，还命人专门制作玉器。北京故宫博物院收藏了上万件玉器，大多是乾隆时期的收藏。北京著名的琉璃厂古玩街、廊房二条玉器街就形成于乾隆年间。

• 镂雕熊形谷纹钺佩（汉）
Openwork Bear-shaped Jade *Yue* Pendant with Millet Pattern (Han Dynasty, 206 B.C.-220 A.D.)

The Collection of Antique Jade Wares

Since ancient times there has been a tradition in China to worship and collect jade: as early as the Xia, Shang, Zhou dynasties, some people of the higher social rank wore jade pendants when they were alive, and were buried with jade when they were dead. In the Han Dynasty (206 B.C.-220 A.D.), a tremendous amount of exquisite jade articles were made, which were adored and

• 定陵出土的花丝镂空金盒玉盂（明）
Carved Golden Box and Jade Bowl Unearthed at *Dingling* Mausoleum (Ming Dynasty, 1368-1644)

collected by the ruling class. From the Three Kingdoms Period to the Northern and Southern dynasties, due to the unrest in society, jade production was stagnated, and so was the collection. Jade production did not recover till the Sui and Tang dynasties when jade pendants, decorations and household utensils, full of life, emerged for the public use, and then jade articles began the process of secularization.

The secularization of jade articles intensified in the Song Dynasty (960-1279), which, together with the emergence of epigraphy (a class of archeology in traditional Chinese culture, the main study objects of which are the bronze articles and inscribed steles of the past generations) promoted the appreciation and collection of jade. Collection had then become a common practice: not only the court produced and collected jade, but also the public made transactions of jade articles.

There was a boom in the development of jade articles in the Ming Dynasty (1368-1644): jade was produced and collected on an unprecedented scale, of which we could catch a glimpse from a tremendous amount of unearthed jade articles of Ming Dynasty in *Dingling* Mausoleum, Beijing. A large number of antique shops and markets selling jade emerged in the city, and there also appeared exquisitely made fake antique jades.

In the Qing Dynasty (1616-1911) jade production reached its peak. Emperor Qianlong, addicted to collecting jade, did not hesitate to spend a lot of human and financial resources, sending his servants across the country to collect jade articles for him and ordering special productions. Beijing Palace Museum collects tens of thousands of pieces of jade, most of which were collected during the Qianlong Period. The famous *Liulichang Antique* Street and *Langfang Ertiao* Jade Street in Beijing came into being also in that period.

- 仿古云龙纹玉洗（清）
Imitation Jade Washer with Dragon and Cloud Design (Qing Dynasty, 1616-1911)

- 白玉牺尊（清）
White Jade Sacrifice Figurine (Qing Dynasty, 1616-1911)

- 玉璜（新石器时代）
Jade *Huang* (Neolithic Period, 4500-8500 years ago)

- 勾云形器（新石器时代）
Curved Cloud-shaped Jade Article (Neolithic Period, 4500-8500 years ago)

居多，其器形与同类的石器相近。常见的有璧、琮、璜、圭等玉礼器；铲、斧、锛、刀等玉工具；勾云形、管形、带钩形、锥形、丫形、环、镯、串饰等装饰玉。

商代早期玉器主要是玉礼器、玉工具，以及少量的装饰品。商代中晚期玉器品种开始丰富起来，有礼器、工具、装饰品、艺术品等。商代玉器器形主要是扁平片状和圆雕，一些片状玉器带有齿牙，是商代玉器特有的特征。西周玉器分为璧、琮、璜等礼

products were made in China. Jade articles of different eras have a great variety of shapes. The shape is an important factor in determining the age of a jade article.

The majority of Neolithic jades are in the shape of a sheet or a circle; their shapes are the same with those of stone wares. Jade ritual objects such as *Bi* (round and plate-shaped jade ritual object with a square-round hole in the center), *Cong* (ritual object, round inside and square outside), *Huang* (semi-annular jade pendant), and *Gui* (belt-shaped ritual object with the bottom being flat and straight, and the upper part being triangle-shaped or straight), jade utensils such as shovel, ax, adz, and knife, jade adornments such as curved cloud-shaped, tube-shaped, hook-shaped, cone-shaped articles and rings, bracelets, decorative strings are all the most often seen. In the early Shang Dynasty jade articles mainly included ritual objects, tools, and a few decorations. In the middle and late Shang Dynasty the forms of jade articles were much enriched. There were ritual objects, tools, ornaments, and works of art. Jade articles in the Shang Dynasty are mainly in the shape of a flat sheet or round sculptures. Some sheet-shape jade

• 玉琮（商）
Jade *Cong* (Shang Dynasty, 1600 B.C.-1046 B.C.)

器、鱼、鹿、鸟等动物玉雕，以及串饰、佩饰等装饰玉。

春秋战国时期，玉礼器减少，玉佩饰增加。其中，玉带钩盛行，数量较多，式样各异。动物形玉器多呈"C"或"S"形。

articles have teeth, and that is a unique characteristic of jade articles in the Shang Dynasty. Jade ritual objects such as *Bi*, *Cong* and *Huang*, jade sculptures such as fish, deer and bird, as well as decorative strings and pendants are the main jade articles in the Western Zhou Dynasty (1046 B.C.-771 B.C.).

In the Spring and Autumn Period and the Warring States Period(770 B.C.-221 B.C.), there were less jade ritual objects but more ornaments. Among them, the jade belt hooks were prevalent in large quantities and different styles. Animal-shaped jades are mostly in C or S shape.

In the Han Dynasty (206 B.C.-220 A.D.), a wide range of shapes of jade appeared. There were not only whole

• 虎纹玉冲牙（春秋）
Jade *Chongya* (Same with *Xi*, an Awl, a Tool Designed to Untie the Knots) with Tiger Pattern (Spring and Autumn Period, 770 B.C.-476 B.C.)

• 玉牛形调色器（西周）
Ox-shaped Jade Color Device (Western Zhou, 1046 B.C.-771 B.C.)

• 白玉兽面纹带钩（战国）
Animal Mask White Jade Hook (Warring States Period, 475 B.C.-221 B.C.)

汉代玉器器形繁多，不仅出现了成套的玉剑饰，还有玉衣、玉握等葬玉，玉舞人、玉刚卯、玉翁仲、鸡心佩等新玉器造型。

• 玉舞人佩（汉）
Jade Dancer Pendant (Han Dynasty, 206 B.C.-220 A.D.)

sets of jade sword adornments, but also funerary jade articles such as jade suit, jade *Wo*, and jade dancer, jade *Gangmao*, jade *Wengzhong*, (*Wengzhong*, a famous warrior in the Qin Dynasty, the original prototype of the jade figure statue), heart shaped pendant and other articles in new shapes.

In the Tang Dynasty (618-907), jade ornaments completely replaced jade ritual objects, and jade utensils were very common. Influenced by foreign culture, there appeared jade articles with the image of barbarians and with the theme of *Feitian*. In the Song Dynasty (960-1279), there appeared jade wine bottles and jade stoves in imitation of bronze articles of the three generations (Xia, Shang and Zhou dynasties). Jade articles for everyday use were mainly small adornments and those for practical purposes. Decorative jade sculptures with images of boy and Buddhist Arhat, as well as flower and bird-shaped jade articles are the most common ones. Jade articles of the Liao Dynasty (907-1125), the Jin Dynasty (1115-1234), and the Yuan Dynasty (1206-1368) have strong ethnic characteristics; Spring-water jade (*Chunshui* jade) and Autumn-hill jade (*Qiushan* jade) are the most influential ones.

• 龙形玉环（汉）
Dragon-shaped Jade Ring (Han Dynasty, 206 B.C.-220 A.D.)

唐代时，玉礼器转化为纯粹的装饰品，玉器皿较多见，另外受外来文化的影响，出现了胡人、飞天等造型的玉器。宋代玉器出现了玉樽、玉炉等仿三代（夏、商、周）青铜器的造型，日常用玉多为小型装饰品和实用品，装饰性玉雕中比

• 云鹤纹玉饰件（宋）
Jade Ornament with Cloud-and-crane Pattern (Song Dynasty, 960-1279)

In the Ming Dynasty (1368-1644), jade articles mainly included ritual objects used in the court, jade wares, jade pendants, jade ornaments and practical devices, among which the jade wares and jade sets of ornaments are the most unique. At the same time, under the impact of men of letters, jade stationery became very popular. The Qing Dynasty (1616-1911) produced the most abundant forms of jade articles. Jade wares, jade pendants, jade stationery, jade ornaments, and jade *Shanzi*: they had all there was to have.

Patterns: Patterns on jade can reflect some characteristics of the times to different extents. Decorations on jade articles of different ages have a certain difference in compositions, shapes and themes, thus the decoration count as an important characteristic of the times.

Before the Shang Dynasty (1600 B.C.-1046 B.C.), jade articles were generally plain without any patterns, or only had some simple incised lines or animal mask designs, etc.; in the Shang and Zhou dynasties, there appeared dragon, *Panchi*, *Taotie*, Cloud and Thunder designs, as well as some fairy bird-shaped decorations. In the Spring and Autumn Period and the Warring States Period (770 B.C.-221 B.C.),

青玉龟巢荷叶佩（金）
Gray Jade Pendant with Tortoise Nest and Lotus Leaves Design (Jin Dynasty, 1115-1234)

较常见的是童子、罗汉形象，以及花鸟形玉器。辽、金、元的玉器带有浓厚的民族特色，影响较大的是"春水玉""秋山玉"。

明代玉器主要有宫廷礼仪用玉、玉器皿、玉佩饰、玉摆件和实用玉器，其中以玉器皿和玉组佩最具特色。同时，受文人墨客的影响，玉文房用具十分盛行。清代玉器器形种类

decorations were gradually enriched; in the Tang Dynasty (618-907), by drawing lessons from the line drawing method, there appeared designs such as *Feitian*, interlocking flowers, which were all depicted exquisitely; In the Song and Yuan dynasties, jade articles had the most colorful decorations, among which the most often seen were dragon, phoenix, and *Chi* designs, and the studs pattern. Characters began to appear on jade in the Ming Dynasty (1368-1644); other common decorations are cloud pattern, dragon design, interlocking flower design, landscape motif, and human image. In the Qing Dynasty (1616-1911) jade decorations were mainly auspicious pictures with special messages, and there were also imperial poetry, inscriptions, and auspicious words used as decorations on jade as well as decorations in imitation

碧玉"莲叶"铭随形砚（明）
Lotus Leaf Design Inscription Jasper Inksone (Ming Dynasty, 1368-1644)

青玉叶形洗（明）
Leaf-shaped Gray Jade Ink Washer (Ming Dynasty, 1368-1644)

最丰富，玉器皿、玉佩饰、玉文具、玉摆件、玉山子等应有尽有。

纹饰：玉器上的纹饰能不同程度地反映出一定的时代特征。不同时代纹饰的构图、造型和主题等都有着较大的区别。所以，纹饰也是玉器的一种重要的时代特征。

商代之前的玉器一般朴素无纹，或只有简单的阴刻线纹、兽面纹等。商周时期，出现了龙纹、蟠螭纹、饕餮纹、云雷纹等，还有一些神鸟纹。春秋战国时期，纹饰渐渐丰富起来；唐代，借鉴线描画法，出现了飞天、缠枝花卉等图案，刻画精细。宋元玉器纹饰丰富多彩，以龙纹、凤纹、螭纹、乳钉纹等图案较为常见。明代玉器上出现了文字，其他较常见的纹饰有云纹、龙纹、缠枝花卉、山水人物等。清代玉器上的纹饰以富有寓意的吉祥图案为主，还出现了御制诗、铭文、吉祥文字等，另外还有仿古纹，以及鸟、兽、鱼、虫等纹饰。

工艺：中国玉器制作的历史悠久，每个时期的玉器加工工艺和艺术风格都不相同。因此，可以通过制玉工艺留下的特征来鉴别玉器的年代。

to antique designs, such as bird, animal, fish and insect.

Craft: China has a long history of producing jade. The crafts and artistic types of different eras are different, thus we can identify the era of jade articles by the characteristics of the craft.

In the Neolithic Age (4500-8500 years ago), people used the method they made stone wares to make jade articles. Round jade articles were not very round, and the thickness was not uniform. There were traces of grinding left on the articles, which were not polished.

In the Shang and Zhou dynasties, people mastered techniques such as hook carving, drilling, pipe drilling, and polishing. In the Spring and Autumn Period and the Warring States Period (770 B.C.-221 B.C.), people began to use abrasive sands to carve and polish jade, and the technical process became more exacting. The techniques were also more advanced than in the Shang Dynasty and the Zhou Dynasty. There were some innovations in the jade production methods in the Han Dynasty (206 B.C.-220 A.D.), of which *Hanbadao* (referring to the craftsmanship of shaping an object with several cuts) is the most distinctive.

新石器时代的玉器制作工艺同石器一样，采用磨制，圆形玉器不是很圆，器壁的厚度也不均匀，器表常留有磨制的痕迹，没有抛光。

商周玉器的制作掌握了勾撤、钻、管钻、抛光等技术。春秋战国时期，开始使用解玉砂琢磨玉器，工艺流程更严格，技法比商周更进步。汉代玉器制作手法有所创新，最有特色的是"汉八刀"工艺。

唐代玉器的制作工艺达到了鼎盛，擅于用较密集的阴线装饰细部，类似绘画上的线描法，有的隐起注重起伏，线条浑厚自然，气韵生动。宋元时期，受不同民族文化的影响，玉器的加工工艺也体现了不同民族和地方的特色，但雕刻线条纤弱，不及唐代玉器生动，器表面往往留有钻痕和砣痕，抛光不是很讲究。

明代出现了浮雕、镂雕工艺，还有双层、三层的镂雕。大件器物的表面往往留有钻痕和砣痕。清代玉器的加工十分讲究，器物的内膛、侧壁等次要部位都处理得一丝不苟。

In the Tang Dynasty (618-907), jades production craft reached its peak. Craftsmen were good at using a more intensive negative scribing to decorate the details, and that was similar to the line drawing method in painting. Some focused on the ups and downs but in a hidden way to make the lines deep, natural, and vivid. In the Song and the Yuan dynasties, influenced by the different ethnic cultures, jade processing technology also reflected the different ethnic and local characteristics, but the lines of the carving were too weak, and less vivid than those in the Tang Dynasty. Moreover, there were always traces of drilling and cutting, and the craftsmen did not pay much attention to polishing.

In the Ming Dynasty (1368-1644), there appeared relief and chasing techniques as well as double and even triple chasing, but there were still traces of drilling and cutting on the surface of large objects. In the Qing Dynasty (1616-1911), lapidary craft was so advanced that minor parts such as the inner and profile parts were handled meticulously.

仿古玉的特征

仿古玉是通过模仿古玉而加工碾磨的玉器，其大致又可分为两种：一种是以低档玉料或替代材料仿制古代玉器的器形、装饰，并用"沁色"伪造时代特征，充当古玉出售，这类仿古玉又称"赝品玉"，多为民间制品。另一种是纯粹从艺术鉴赏的角度来仿造，力求在器物的整体风格上模仿古玉。

仿古玉兴起于宋代。因为宋代时厚古之风盛行，古玉受到官宦文人的追捧，仿古玉应运而生。宋代仿古玉有宫廷制和民间制之分。宫廷制的仿古玉具有较高的艺术价值。

The Characteristics of Imitative Antique Jade Pieces

Imitative Antique Jade articles are carved and grinded in imitation of antique jade, of which there are mainly two types. One is to process low-grade jade material or alternative materials in imitation of shapes and decorations of antiques, and use artificial "color impregnation" to counterfeit characteristics of the time and they sell these imitations as antiques. These imitations were called fakes, and are mainly made by ordinary craftsmen. The other is to completely imitate antiques' style only for aesthetic purpose.

People began to counterfeit antique jade in the Song Dynasty (960-1279). As the trend to adore the ancient was prevalent in the Song Dynasty, antique jade was sought after by the elite government officials and men of letters, thus fake antique jades came into being with the tide of fashion. The antique jades had the difference of being civil-made or court-made. Fake antique jade made by the court system has a relatively high artistic value. At that time the emergence of fake antique jades promoted the trend of "back to the ancients", and the demand for fakes surged so high that

● 沁色双龙纹玉镯（宋）
Double Dragon Design Color Impregnated Jade Bracelet (Song Dynasty, 960-1279)

● 仿古兽耳云龙纹玉簋（宋）
Imitative Antique Animal Ear Jade Bowl with Cloud and Dragon Design (Song Dynasty, 960-1279)

当时仿古玉的出现引起了复古风潮，因而需求量剧增，也导致民间商人通过制作赝品玉来获利。宋代时仿古玉已出现伪沁，方法多模拟墓葬的环境，然后用火烤、染色等方法，加速沁色过程，达到局部改变仿古玉的表面颜色和质感的目的。

明代仿古玉的制作手段更加丰富，有的只仿古代玉器造型，纹饰却带有鲜明的本朝特征；有的造型、纹饰都明显带有本朝特征，却

private traders began to make profit by making fake antiques. In the Song Dynasty(960-1279), there already appeared fake color impregnation: a simulation of burial environment, then with burning, staining and other methods to accelerate the impregnation process in order to partially change the surface color and texture of the fake antique jades.

The methods to counterfeit antiques were largely enriched in the Ming Dynasty (1368-1644). Some only imitated the shapes of antiques, and kept the distinctive characteristics of their own era in the decorations, but used mutilation and artificial color impregnation to counterfeit; others imitated both the shape and the decoration of antiques, and by using various methods succeeded in making their fakes fool even an expert. In the Ming Dynasty(1368-1644), people used the method of burning and dying to imitate the color impregnation of antique jades, and they used colors such as earth

● 仿古簋式玉炉（明）
Imitative Antique *Gui*-shaped Jade Stove (Ming Dynasty, 1368-1644)

• 仿古蟠龙纹玉觥（明）
Imitative Antique Jade Bottle with the Dragon Pan Design (Ming Dynasty, 1368-1644)

用致残、伪沁等手段来造伪；还有的造型和纹饰均仿古玉，并成功地运用各种作旧手段达到真假难辨的程度。明代时使用烧染的方法仿制古玉器的沁色，有土绣、血浸、黑绣、铜色等。

清代仿古玉制作已经达到极高的境界，宫廷仿古玉或按照古玉图录仿制，或在图录基础上稍加变化，以质朴浑厚、古色古香的特有格调为后人所推崇。清代仿古玉不断尝试新的作旧手段，注重人工仿

rust, blood red, black rust and bronze.

In the Qing Dynasty (1616-1911), Imitative antique jade production reached a very high state; the court either imitated antiques in accordance with the pictures of antique jades or made some little changes on the basis of antiques' pictures. Those imitations, simple and modest, with a unique quaint and archaic flavor, are highly praised by later generations. Craftsmen in the Qing Dynasty(1616-1911) had constantly tried new methods to counterfeit antiques, focusing on artificial color impregnation and mutilation process so that they made it difficult to distinguish between the authentic and the fake

• 仿古沁色玉薰（清）
Color Impregnated Imitative Antique Jade *Xun* (Qing Dynasty, 1616-1911)

• 仿古沁色素面玉琮（清）
Color Impregnated Imitative Antique Jade *Cong* with Plain Surface (Qing Dynasty, 1616-1911)

沁色和人工致残工艺，仿制水平极高，真假难辨。据文献记载，清代沁色手法有鸡骨白、羊玉、狗玉、梅玉、风玉、叩锈、提油法等。

古玉作伪手法

自仿古玉出现以后，古玉作伪手法就层出不穷。古玉作伪手法主要是做旧和人工染色两种。做旧手法又分为在古玉上做改造和人工致残两种；人工染色手法则是各式各样，只要古玉沁色中有的，造假者总能通过各种方法制造出一样的沁色效果来。

because of their advanced techniques. According to historical records, the methods to impregnate colors in the Qing Dynasty(1616-1911) are mainly chicken bone whitening (by burning the jade), the sheep method (by planting the jade into a living sheep's leg), the dog method (by planting the jade into the stomach of a newly dead dog whose blood has not coagulated yet), the plum method (by boiling the jade in the plum juice), the wind method (by exposing the jade in a snow storm weather after boiling it in the plum juice), the rust method (by eroding the jade with iron rust), oil extraction method (by immersing the jade in a kind of grass juice, and then burning it).

Techniques to Counterfeit Antiques

Since the emergence of imitative antique jade, the methods to make these imitations had sprung up in an endless stream. Making jade look old and artificial coloring are the two main ways of imitation. Old jade modification and artificial mutilation are the two methods to make jade look old, and there are a wide range of artificial coloring techniques: counterfeiters could always find a way to produce the same color impregnation effects with those of the authentic ones.

古玉做旧手法

旧玉改造：指后人对前代传世古玉进行改制。用于改制的旧玉，一般器物破损或做工粗糙，进行改造后的玉器不易被识破。还有的是用古代余料，或是前人未制完的半成品，依料借形，加以改造，使之成为一件当时流行的器物。

旧玉后动刀：指在古玉上增添纹饰。这种玉器所用的玉料是真的，器形也是古时式样，只有纹饰是后人增添的。

砣碾致残：指在仿古玉器表面顶出圆形或椭圆形、深浅不一、长短不齐的点坑或线条，粗看似长期磕碰所致，具有传世古玉的面貌。

敲击致残：指利用玉料有绺（玉器表面的褶皱）的性质，轻轻敲打仿古玉器的器身，一般力度控制在使小绺变大绺、大绺变裂痕时为宜。为了掩饰裂纹的新痕，还要在这些部位进行染色处理，有的甚至将仿古玉器的器身故意断裂。

- 带牛毛纹的白玉尊【局部】
White Jade *Zun* with Ox Hair Pattern [Part]

砂磋毛道：指在抛光时有意保留一部分糙面，或者抛光后故意用细砂子在器表稍加磨磋，可做出或多或少的纤细划痕，粗看像是传世时因手摩挲而形成的痕迹。

仿牛毛纹：牛毛纹是古玉器表面上出现细密如牛毛状的裂隙，是古玉器的一种外表特征。伪造牛毛纹的方法是把白玉放在寒冷的地方，靠自然的热胀冷缩，使玉器的表面产生细密如牛毛状的裂纹。

Methods to Make Jade Look Old

Old Jade Modification: It refers to the changes made by later generations on old jade handed down from previous generations. The old jades for modification were usually rough works or had been damaged. After being modified, these jades cannot be seen through easily. Some others use left-over ancient materials or semi-finished products of previous generations, and according to the particular shape of the piece, transform it to a very popular object.

Adding Patterns to the Old Jade: It refers to adding decorations to the old jade. For this kind of jade articles, the jade material used is authentic, the shape is the ancient style, and only the decoration is added by later generations.

Mutilation by Cutting and Grinding: It refers to creating circular or oval pits, dots and lines of different shades and lengths on the surface of the jade which look like the damages caused by a long history of bumping at the first glance, thus the imitative jade has an appearance of an antique handed down from the ancient times.

Mutilation by Percussion: It refers to gently tapping the body of the imitation, and as there are the folds on the jade surface, using the appropriate force to make the little folds become large ones, and large folds cracks. In order to cover up the new marks, the dyeing process is needed, and some even deliberately break the imitation in the middle to make it look ancient.

Roughening Method: It refers to keeping a part of the jade article unpolished, or deliberately roughening the jade surface with fine sands to make some fine scratches on the jade. At first glance these rough characteristics are like traces left by many hand caresses during the jade's collection history.

Imitating the Ox Hair Pattern: The ox hair pattern on the surface of antique jade is actually the fine and delicate little cracks, and it is a kind of physical feature of antique jade. To imitate the ox hair pattern, one puts a white jade in a cold place where the natural expansion and contraction would create fine and delicate ox hair like cracks on the jade surface.

人工染色手法

沁色是玉的一种瑕疵、缺陷，但作为古玉的外观标志，也具有一定的美感，对于鉴别玉器有十分重要的作用，因此针对不同的沁色出现了不同的人工染色手法。

提油法：方法是将玉器需染色的部分放入油锅中炸，玉器表面的颜色会因炸的时间长短而出深浅不一的枣皮红、橘皮红等颜色。

酸蚀法：将玉器放入强酸中腐蚀，使玉器表面呈现出高低不平的现象，取出后加热着色。

仿鸡骨白：鸡骨白是一种天然沁色，似鸡的骨头一样，不透亮。浙江余杭一带的良渚文化玉器大部分被蚀成鸡骨白。伪造鸡骨白沁色的方法是将新玉用炭火煨烤，在炭火未冷却时用凉水泼在玉上，取出后与受地火影响的古玉相似，呈鸡骨白颜色。

仿血沁：将玉埋入猪血和黄土合成的泥中，埋入大缸，时间久了就会有土沁、黄土绣、血沁等痕迹。还有把玉放在狗血或羊血中浸泡，形成血沁的效果。

仿水银沁：水银沁是指玉上沁色黑如乌金。伪造水银沁的方法是把成型的玉器中想保留原有玉质的部位用石膏粉贴住，然后用黑色的乌木屑或暗红色的红木屑煨烤，这样除了石膏贴住的部分，其他地方都能沁上颜色，其效果与水银沁的古玉相似。

仿土沁：土沁是指古玉上带有黄斑。伪造土沁的方法是把胶水涂在玉器上，然后埋于黄土中，使玉器表面沾满黄土斑。埋得时间越长，其效果就越真实。

Artificial Coloring Techniques

Color impregnation is a kind of flaw and defect for jade articles, but as a physical feature of antique jade, it also has a certain beauty. For the identification of authentic jade it plays a very important role. In order to imitate different kinds of color impregnation, there emerged different artificial coloring techniques.

• 水银沁青白玉卧鹿（明）
Greenish-white Jade Display with a Lying Deer Design with Mercury Impregnation (Ming Dynasty, 1368-1644)

Oil Extraction Method: It is to put the part of jade in need of dyeing into a boiling oil pot. According to the time the different parts of the jade surface are fried, they will have different shades of date peel red and orange peel red.

Acid Etching Method: It is to put the jade into the strong acid to corrode it so that the surface of the jade will be uneven, and then after having heated it, color it.

Imitative Chicken-Bone White: Chicken bone white is a natural color impregnation; it is called so as the jade's color is like chicken bones, white but not transparent. Jades of Liangzhu Culture (4000-5000 years ago) in Yuhang area of Zhejiang Province were largely impregnated into chicken bone white. To imitate chicken bone white impregnation, people simmer the new jade with charcoal fire and splash cold water onto the jade when the charcoal has not cooled. The new jade processed in this way, like the antique jade influenced by the earth fire, shows a chicken bone white color.

Imitative Blood Impregnation: To imitate blood impregnation, one buries the jade into the mud of pig blood and yellow earth, and then stores it in a vat. After a long time, there will be traces of yellow earth rust, soil and earth impregnation. There are also people who soak the jade in dog or sheep blood to imitate the blood impregnation effect.

Imitative Mercury Impregnation: It refers to brilliant black color on the jade. To imitate mercury impregnation, one covers the part of the jade which does not need dyeing with the gypsum powder, then simmers the jade with fire made from black wood or dark red wood flour, thus except for the part covered by gypsum powder, the jade will have the color similar to the mercury impregnation of antiques.

Imitative Soil Impregnation: It refers to the yellow spots on jade. To imitate soil impregnation, one applies the glue to the jade, and then buries it in yellow earth in order to make the surface of jade covered with yellow earth spots. The longer the jade is buried, the more real the effect is.

> 如何选购现代玉

古玉流传较少，而且价格高昂，由此，购买现代玉器也是不错的选择。尤其是在玉器制作工艺成熟精湛的基础上，现代玉器不仅继承了中国传统玉文化的精髓，更涌现出许多新题材、新形式的作品。而且选购现代玉器不需要特别考虑年代或人工作伪的痕迹，只需要注重鉴别玉料和雕工两方面的品质。

现代玉器的特点

从20世纪50年代至20世纪80年代，中国玉器行业加强了对艺术品工厂体制的管理，推动了玉器行业的发展，提高了玉器制作的水平。在此期间，赝品古玉的生产基本被禁止了，不存在以次充好的问题，货真价实。这一时期主要的玉器厂

> How to Buy a Modern Jade Article

There are not many antique jades in the market, and they are all very expensive, therefore, to buy modern jades is also a good choice. Moreover, on the basis of advanced and exquisite jade production craft, modern jade articles not only inherits the essence of traditional Chinese jade culture, but also includes many works of new themes and new forms. Besides, to buy a modern jade, one does not need to consider the age of the jade or traces of artificial imitation; he only has to focus on the quality of the material and the carving.

Characteristics of Modern Jades

From the 1950s to the 1980s, China's jade industry enhanced the managenment of the art factory system, which promoted the whole industry and improved the

• 和田白玉带皮"喜获大礼"摆件（现代）
Cover-retained White Kohtan Jade Ornament with the Theme of "Happily Receiving a Magnificent Gift" (Modern Times)

• 翡翠摆件"华夏古风"（现代）
Jadeite Ornament with the Theme of "Ancient China Landscape" (Modern Times)

培养了一批艺人，他们中大多数人成为当代著名的玉器制作大师，由他们设计制作或领衔制作的作品，无论是工艺技术，还是艺术价值都很高。

在制作工艺方面，由于工具设备得到改进，尤其是大型的电动工具和小巧好用的软轴钻的出现，使得现代玉器加工可以施以更高级的工艺，所以大型玉山、玉人等难度高的玉雕名作时有问世。由于原料珍贵，在制作的设计环节，玉雕题

level of jade production. During this period, the production of imitation antique jade was basically put to an end, and there was no cheating the customers. The goods were genuine and the price was fair. In this period the main factories trained a group of jade artists, most of whom became well-known contemporary masters of jade production. Their works and works made by a team lead by them have fine workmanship as well as high artistic value.

Since the tools and equipments have got improved, and especially for the emergence of large motor-driven tools and small and handy flexible shaft drills,

材的选择往往优先以玉色为指导，巧用天然色料，精心设计出最能表现玉质美和玉色美的题材内容，甚至是有缺点的玉料，也被能工巧匠们灵活设计，充分利用玉料本身的特点，化瑕为瑜。

现代玉器的选购

现代玉器的种类不像古玉的种类那么庞杂，但也可大致分为玉雕工艺品、玉饰品、玉保健器材三大类。

玉雕工艺品：主要包括玉器皿、玉摆件、玉插屏等。现代玉

• 白玉东方巨龙花熏（现代）
White Jade Flower *Xun* (a Kind of Incense Burner) with the Oriental Dragon Design (Modern Times)

more advanced technology can be applied to modern jade processing. Masterpieces in a high degree of difficulty, such as the large set of jade mountains, and jade human statues are often presented to the public. As the raw materials are very precious, the choice of the theme is usually guided by the quality of the raw jade. Artists make a brilliant use of the natural color of the raw material and in a careful and delicate way design a subject matter which can best bring out the beauty of the texture and color of the jade. Even a jade material with a flaw can become a precious article after being carefully designed and processed according to its own characteristics by a skillful craftsman.

How to Buy Modern Jade Items

Unlike the antique jades, modern jades may not have that many complex types, but they can also be broadly divided into three categories: jade carving works, jade accessories, and jade health-care equipments.

Jade Carving Works: It includes jade wares, jade ornaments, jade table screens, etc. The value of a modern jade is determined by the raw material, workmanship, originality, reputation of the artist and so on. Material and

雕工艺品的价值由用料、雕工、创意、大师名气等因素组成。用料和雕工是玉器的成本价值，创意和大师名气则属于玉器的附加价值。

现代玉器比较注重设计者和制作者的名号，很多玉雕大师的作品已经成为现代玉石工艺品中的"名牌"。一方面，玉雕大师的作品一般用料珍贵，雕工精湛，是精品中的精品；另一方面，一位玉雕大师的创作黄金期为30～55岁，每年所创造的作品为8～10件，一位大师的黄金期作品不过200件左右。玉雕大师的作品具备了稀缺性，保证了其在未来的升值空间。

workmanship are the cost-based value of the jade article, while originality and the artist's fame are the added value.

Collectors of modern jade pay much attention to reputations of the designers and the craftsmen. Works of many jade carving masters have become famous brands in modern jade markets. On the one hand, the raw materials of jade works by masters are generally very precious; with the exquisite carving, they can be counted as the finest among masterpieces. On the other hand, a master's golden period for creation is when they are 30 to 55 years old; they create about eight to ten works each year; therefore, a master's golden period works cannot be more than 200 pieces or so. Jade works of masters have the rarity to ensure their potential for appreciation in the future.

The number of jade ornaments is the largest in jade works. Jade ornaments can be collections as well as furnishings, and they should be coordinated with the decoration style and color tones of the display space. Out of consideration for security of the jade ornaments, there is no need to pursue expensive high-end jade carvings. Some practical jade articles such as jade tea sets, jade wine sets, and so on could also be used for display.

- 羊脂玉霄汉回翔玉雕（现代）
Mutton-tallow Jade Display with the Theme of Eagle Hovering above the Sky (Modern Times)

● 翡翠茶具
Jadeite Tea Set

玉摆件是玉雕工艺品中数量最多的，也可用作收藏或陈设。玉摆件与陈设空间的装修风格和色调要相协调。出于玉摆件陈设的安全性考虑，不必刻意追求高档昂贵的玉雕，有些实用玉器也可用于陈设，如玉茶具、玉酒具等。

现代玉饰品的种类有项链、戒指、手镯、耳饰、玉佩、胸花等。玉饰品的用料、设计、雕刻都是其价值的重要因素。选购玉饰品时也要考虑佩戴者的年龄、穿着、身材、脸型等因素。

项链是现代玉饰品的一个主要类别，常用的材质有和田玉、翡翠、蓝田玉、独山玉、梅花玉，还包括萤石、孔雀石、玛瑙、琥珀等

Modern jade accessories include necklaces, rings, bracelets, earrings, pendants, brooches and so on. The materials, designs, and carvings are all important factors to determine their value. When buying jade accessories, one has to consider the wearer's age, dressing style, body shape, face and other factors.

Jade necklace is one of the main types of jade accessories. The commonly used materials are Khotan jade, jadeite, *Lantian* jade, *Dushan* jade, Plum-blossom jade, as well as fluorite, malachite, agate, amber, and other special types of jade. There are few fake jade necklaces, and most of the necklaces have certificates of authenticity issued by the authorities.

Jade ring is another main category of jade accessories. Jade rings are generally divided into two types: the stirrup-shaped ring style and inlaid ring gem style. Stirrup-shaped ring style is cut and grinded from a single piece of raw material. The ideal one should have a uniform color, a fine and even texture, which is usually made from jade materials that are easy to process, such as agate and *Xiu* jade. For the inlaid ring gem style, only the ring gem is made from jade, and is wrapped by all kinds of metal brackets. The common shapes of the ring gem are

• 翡翠项链（现代）
Jadeite Necklace (Modern Times)

round, oval, saddle, olive, water drop, etc. The materials include chatoyant opal, purple crystal, turquoise, malachite, lapis lazuli, jadeite, etc. The most popular is the jadeite ring gem. Those elegant, long "water head" ones, which have a pure and even green color and a weight above 5 carats (1 gram), are the best.

Jade bracelet is of the mainstream of the jade accessories, of which the main materials are *Xiu* Jade, jadeite, Khotan jade, *Dushan* jade, agate, quartzite jade, Plum-blossom jade and so on. Jade bracelets usually have a round shape, and there are also some square, oval, circular flat bar-shaped bracelets. Modern jade bracelets have a fine polishing and plain inner ring, and usually have no decorative patterns, which brings out their natural color tone and the beauty of clearness and smoothness. However, with the increase of people's aesthetic requirement, there are more and more bracelets with decorative patterns, such as the line-patterned, carved-patterned, and twisted-wire-patterned bracelets, appear on the market. One should choose a bracelet according to the length and shape of the human body. Thin and short people should choose a thin bracelet, while plump and tall ones should choose

特殊玉种。玉质项链作伪者较少，且多附有权威机构的鉴定证书。

戒指是玉饰品的第二个大类，一般可分为马镫型与镶嵌戒面型两种类型。马镫型戒指是指整个戒指都是由玉料一体切磨而成的，理想中的马镫型戒指颜色均匀，质地细

• 翡翠镶嵌戒面型戒指（现代）
Inlaid Jadeite Ring (Modern Times)

• 翡翠马镫形戒指（现代）
Stirrup-shaped Jadeite Ring (Modern Times)

a slightly thick bracelet.

Materials for earrings include jade, agate, turquoise, malachite, amber, jadeite, and so on. Earrings made of jade can be divided into two types: studs and eardrops. Jade earrings have different shapes and sizes. One should choose the earrings according to the body and face shapes. Those of a short stature should choose studs in order to look more

腻，这类戒指的材料多用玛瑙、岫玉等易加工的玉料。镶嵌戒面型戒指只有戒面是玉石，戒面被各种金属托架包裹，常见的戒面形制有圆形、椭圆形、马鞍形、橄榄形、水滴形等，材质包括变彩蛋白石、紫晶、绿松石、孔雀石、青金石、翡翠等。其中最受欢迎的是翡翠戒面，以饱满大方、绿色纯正均匀、水头好、重量在5克拉以上者为佳。

手镯也是玉饰品中的主流玉饰，材质主要有岫玉、翡翠、和田玉、独山玉、玛瑙、石英岩玉石、梅花玉等。其制式多为圆形镯，也有部分方形镯、椭圆形镯、扁条镯等。现代玉质手镯抛光很好，通常无纹饰，内圈素面，从而表现玉质天然的色泽和莹润之美。随着人们审美要求的提高，纹饰类手镯如绳

• 翡翠圆形手镯（现代）
Round-shaped Jadeite Bracelet (Modern Times)

• 翡翠椭圆形手镯（现代）
Oval-shaped Jadeite Bracelet (Modern Times)

• 翡翠扁条镯（现代）
Circular Flat Bar-shaped Jadeite Bracelet (Modern Times)

graceful and delicate while the tall ones can wear eardrops to make them more beautiful. Those with a round face should choose water-drop-shaped or oval-shaped slender eardrops, which would make the face look slenderer while those with a long face should choose round, fan or square-shaped ones as eardrops with a horizontal design are fit for them.

Jade pendants include those jade articles hung in front of the breast and those worn on the belt. The most often seen are Jade Avalokitesvara (*Guanyin*), jade Buddha, twelve Chinese zodiac signs, longevity lock (*Changmingsuo*), heart-shaped jade, jade buckle and jade tablets carved with a variety of traditional auspicious patterns, etc. When buying jade pendants, one needs to consider the wearer's age, gender, color and other factors.

纹手镯、雕花手镯、绞丝手镯等也越来越多地出现在市场上。手镯要根据人的高矮胖瘦来选择。瘦矮的人应选择稍微纤细的手镯，胖高的人则应选择稍微粗些的手镯。

耳饰的材质包括玉、玛瑙、绿松石、孔雀石、琥珀、翡翠等。用玉制成的耳饰可以分为耳钉和耳坠两种。玉石耳饰的形状大小各异，选购时应根据体型、脸型来选择耳饰的形状。如身材矮小的可以选择耳钉，显得玲珑秀气；身材高挑的可以佩戴耳坠，增加美感；圆脸可以选择水滴形、长椭圆形等修长型耳坠，可以使脸型显得修长；长脸适合圆形或横向设计的耳饰，可以

• 翡翠镶红宝石耳钉（现代）
Jadeite Earrings with Inlaid Rubies (Modern Times)

选择圆形、扇形或方形的耳坠。

玉佩包括挂于胸前的佩件及佩于腰带之上的玉件，常见的有玉观音、玉佛、十二生肖、长命锁、玉鸡心、玉扣及雕有各种传统吉祥图案的玉牌等。选购玉佩要考虑佩戴者的年龄、性别、肤色等因素。

玉保健器材：古代中国人就已发现了玉的保健作用，其所含的锌、铁、铜、锰、镁、钴、硒、铬、钛、锂、钙、钾、钠等多种微量元素，对人体有益。为迎合人们对健康的需求，现代玉器中出现了专门的玉保健器材，如玉枕、玉石健身球、玉石按摩器等。使用者最好具备一些养生保健常识，以免误用产生副作用。

Jade Health-care Equipments: In ancient times, Chinese had found that jade is good for human health. It contains many microelements including zinc, iron, copper, manganese, magnesium, cobalt, selenium, chromium, titanium, lithium, calcium, potassium, sodium, etc. In order to meet the health-care needs of people, there emerged several special jade health-care equipments such as jade pillow, jade fitness ball, jade massager, etc. Users of these equipments should better possess some basic health-care knowledge to avoid the side effects by misuse.

中国近现代制玉名家
Chinese Contemporary Masters of Jade Production

王树森（1917—1989），北京人，北京玉雕艺术家，北玉四杰（其他三人为潘秉衡、何荣、刘德瀛）之一。他的作品擅长"小中藏大""薄中显厚""平而反鼓"（即凸出来，具有立体感）。

Wang Shusen (1917-1989) is a jade carving artist who is from Beijing. As one of the Four Northern Jade Masters (the other three are Pan Bingheng, He Rong, and Liu Deying), he was good at hiding the large in the small, displaying the thick in the thin, and making the flat protruding (in order to make a three-dimensional effect).

潘秉衡（1912—1970），河北固安人，玉器艺术家，他发明了套料取材法，并恢复发展了压金银丝嵌宝技艺。

Pan Bingheng (1912-1970), a jade artist, is from Gu'an, Hebei Province. He invented the nesting method to obtain jade materials from a raw piece (*Tiaoliaoqucai*), and restored and developed the skills to press gold and silver filaments onto the jade.

刘德瀛（1913—1982），河北霸州市人，北京玉器业老艺人，擅长玉器花卉设计、雕刻。

Liu Deying (1913-1982) is from Bazhou City, Hebei Province. He was good at designing and carving flowers on jade.

何荣（1907—1983）河北香河人，北京玉器业老艺人，曾任北京市玉器厂玉器设计指导，他擅长人物玉雕的制作。

He Rong (1907-1983) is from Xianghe, Hebei Province, one of the Old Jade Artists of Beijing. He used to be in charge of designing and supervising in Beijing Jade Articles Factory, and he specialized in the processing of jade figures.

夏长馨（1912—1989），天津武清人，中国工艺美术大师，擅长平素产品的花纹设计，对各种玉器图案、兽头，深浅浮雕设计得当，自成一派。

Xia Changxin (1912-1989) is from Wuqing District, Tianjin City, a master of Chinese arts and crafts. He was good at designing the patterns for simple and modest works. The depth of relief in his designs for a variety of jade articles and animal heads are so appropriate and excellent that he formed his own school.

王仲元（1913-1994），北京人，中国工艺美术大师，以玉雕花卉闻名，其作品构思巧妙，造型优美大方，擅长使用俏色。

Wang Zhongyuan (1913-1994), a master of Chinese fine arts and crafts, is from Beijing. He is well known for jade carving with flower designs. His works are ingenious, original and elegant in shape. He was good at using pretty and brilliant colors.

李博生（1941— ），师从何荣、王树森，中国工艺美术大师。其作品以形带情，情景交融，素以构思巧妙、题材新颖、造型优美、刻画细致著称。
Li Bosheng (born 1941), learning from He Rong and Wang Shusen, is a master of Chinese fine arts and crafts. His works are capable of expressing the emotion by the form in a harmonious way. His works have been famous for brilliant and original design, novel theme, elegant shape, and detailed characterization.

顾永骏（1942— ），中国工艺美术大师。开拓和完善了扬州明清年代的玉雕名品"山籽雕"，借鉴中国画的表现手法，在玉雕创作中借鉴中国画的构图、线条，风格俊逸洒脱，自成一体。
Gu Yongjun (born 1942) is a master of Chinese fine arts and crafts. He developed and perfected the famous *Shanzi* carving of the Yangzhou city in the Ming and Qing dynasties. He drew lessons from expression techniques of Chinese painting to create jade carvings, using the compositions and lines of Chinese painting. With his free and easy style, he has formed his own school.

冯道明（1947— ），河北雄县人，北京工艺美术大师，其作品注重传神和意境，用料严谨，设计巧妙，形象优美，富有变化。
Feng Daoming (born 1947), a master of fine arts and crafts in Beijing, is from the Xiong County, Hebei Province. His works focus on expressing the mood and the artistic conception. His works, made from strictly-chosen raw materials, and cleverly designed, are full of changes and beautiful images.

宋世义（1942— ）北京人，中国工艺美术大师。他的技法全面，作品题材广泛，构思严谨、工艺细腻，尤其擅长制作俏丽多彩的玛瑙和珍奇多姿的珊瑚。
Song Shiyi (born 1942), a master of Chinese fine arts and crafts, is from Beijing. He masters the comprehensive techniques, and his works have a great variety of themes. His idea is well-conceived while the processing is delicate. He is especially good at making brilliant and colorful agate articles as well as rare and graceful coral articles.

江春源（1947— ），河北省雄县人，中国工艺美术大师，主要擅长花卉、雀鸟、炉瓶、走兽的创作设计。其作品清新雅致，内容常取诗词情景，精雕细琢。
Jiang Chunyuan (born 1947), a master of Chinese fine arts and crafts, is from the Xiong County, Hebei Province. He specializes in designing and making jade flowers, birds, censers, bottles, and animals. His works are fresh and elegant, of which the contents are usually scenes taken from poetry, and they are all exquisitely carved.

吴德升（1961— ）中国玉石雕刻大师。他的作品题材广泛，尤以人物玉雕见长，他把西方雕塑夸张和立体感强的特色融入玉雕创作中，作品具有沉稳大气、工艺精湛、古朴灵动的特点。
Wu Desheng (born 1961) is a Chinese master of jade carving. His works have varied themes, and he is especially good at carving figures. He introduced the exaggeration style and three-dimensional effects of Western sculpture into his own jade creations. His works, calm, poised, and elegant with classic simplicity, are exquisitely crafted.

张伟良（1962— ）中国玉石雕刻大师。其作品选材考究，题材广泛，擅长把玉石的质地、色彩与作品的造型、意境相结合，风格高古雅逸，清丽幽隐。
Zhang Weiliang (born 1962), a Chinese master of jade carving. The materials for his works are carefully selected, and his works have a wide range of topics. He is good at coordinating the natural texture and color of the raw jade with the shape, mood and style of the work. He has a high-end classic style, with an elegant secluded feeling.

姜文斌（1963— ），北京人，中国玉雕大师，其作品严谨规整，线条流畅，轻盈明快，做工精细。
Jiang Wenbin (born 1963), a China jade carving master, is from Beijing. His works are rigorously structured, and have flowing lines. Bright and elegant, his works are exquisitely carved.

高毅进（1964— ）江苏扬州人，中国工艺美术大师。他擅长玉器器皿、仿古、走兽、杂件的设计制作。
Gao Yijin (born 1964), a master of Chinese fine arts and crafts, is from Yangzhou, Jiangsu Province. He specializes in designing and making jade wares, imitation antiques, jade animals, and other miscellaneous articles.

薛春梅（1965— ）江苏扬州人，中国工艺美术大师。她擅长人物玉雕的创作设计，尤其在表现孩童的神态、动态上独具匠心，不落俗套。
Xue Chunmei (born 1965), a master of Chinese fine arts and crafts, is from Yangzhou, Jiangsu Province. She specializes in designing and creating jade figures. She has a unique way to represent children's demeanor and activities, which is creative and unconventional.

附录
Appendix

> 玉器盘玩注意事项

出土的古玉，往往受到锈蚀和色沁，遮盖了玉器本身的质地与色彩，要通过盘玩才能使其稍微恢复本来的光彩。盘玩玉器十分讲究，只有通过科学的盘玩方法才能尽最大可能恢复古玉本来的面目。

对体轻、质地疏松、沁色浓厚的古玉，先用旧白棉布反复擦拭，再用新白棉布擦拭。每擦拭一次都要使玉发热，反复擦拭几次之后，灰尘、浊气会从古玉上渐渐褪去，带有沁色的地方自然凝结、收敛，颜色越敛越艳。

对质地坚硬、沁色浅薄的古玉，可先用水煮法，即将玉悬空挂在装有茶叶末的大瓷罐中，瓷罐中加清水，用文火蒸，将沁入古玉体内的土气剔除。然后，趁热把古

> Some Tips on Jades Treatment

The unearthed antique jades are usually rust corroded and color impregnated; therefore, their own texture and color are covered. Only by the treatment, can their original luster be restored in some degree. The treatment of a jade article is very exacting, and only by using the scientific method, the original appearance of jade can be restored to the maximum extent.

For antique jades with a light weight, a loose texture and a deep color impregnation, one has to first stroke it with an old white cotton repeatedly, and then with a new white cotton cloth. Every time one strokes it, has to make the jade heated. After several times of cleaning it, dusts and the foul air will gradually leave the jade. The impregnated parts of the jade will consolidate and converge naturally, and

玉取出，用细密的棕刷清理。稍凉后再蒸，反复几次后，用新的粗白棉布擦拭，这样可将古玉的土气脱尽，色彩焕然一新。

对玉质坚硬，全身被灰土包裹的古玉，将其放入装有稻壳、木绒草的袋子中，用手擦、揉、搓，使玉恢复本来的面目。

- 鸟形玉饰（新石器时代）
Bird-shaped Jade Ornament (Neolithic Age, 4500-8500 years ago)

- 蟠龙玉环（魏晋南北朝）
Jade Ring with Pan-dragon Design (Wei, Jin, Southern and Northern Dynasties, 220-581)

show a more brilliant color.

For antiques with a hard texture and shallow impregnation, one can use the boiling water method. That is to suspend the jade in a large porcelain jar full of tea dusts, and add pure water to the jar, and then simmer it, thus the earth air impregnated into the antique is removed. As soon as the jade is taken out of the jar, clean it with a fine brown brush. When it becomes cooler, simmer it again. Repeat this process several times and stroke the jade with a coarse new white cotton cloth. When the earth air is completely removed, the antique will take on an entirely new bright color.

For antique jade all wrapped up in grey earth, and with a hard texture, one can put it into a bag full of rice husks and wood velvet grass, and then rub and squeeze it so as to restore the original features of the jade.

For antique jade which has a deep color impregnation but is not wrapped up in grey earth, one can put it into a bag full of bamboo leaves and rice bran shells and rub it. If there appears toad-skin-like patterns on the surface, one has to put it into a big bag full of rice husks and wood velvet grass and do the rubbing. If the color of the jade is still the same after

对沁色较多、无灰土包裹的古玉，将其放入装有竹叶或稻糠壳的袋子中进行盘玩。在盘玩时，如表面出现似蛤蟆皮，要先放入装有稻壳、木绒草的袋子盘玩。如果经上述盘玩之后，古玉的玉色仍不变，还可继续采用稻草灰煮水，将玉悬挂于瓷罐中，用文火蒸煮，待玉色稍变后，马上用以上各种盘玩法盘玩。

对经过盘玩，但沁色干枯的古玉，可先用肥皂水、皂角水煮，再用竹叶、稻糠壳装袋盘玩，这样可使玉体通透细润，沁色亮丽、均匀、不呆滞。盘玩后的玉，要时常用手抚摸擦拭，或随身佩戴，以身体之气滋养玉。

the above treatments, one can continue to suspend the jade into a porcelain jar full of water with straw ashes and simmer it. When the color has slightly changed, stroke it immediately by various ways.

If the color impregnation becomes dry and dull after the treatment, one can first boil the jade with soap water or gleditsia water, and then put it into a bag of bamboo leaves and rice bran shells to rub it, and thus the jade will become fine, smooth and more transparent, with a bright and even color impregnation, and not dull any more. After the treatment, one should often caress the jade by hand or wear it wherever one goes, so that the air emanated from the human body can nourish the jade.

> 玉器保养注意事项

避免碰撞

玉石的硬度虽高，但是受碰撞后很容易裂，特别是古玉的玉质疏松，碰撞后容易破碎。即使是碰撞后没有出现破碎或断裂，也可能在

> Tips on Jades Preservation

Avoid Collision

Although jades are rather hard, they are easily cracked by collisions. Besides, antique jades have a rather loose texture, and they are easily broken into pieces after the collision. Even if the jade is not broken or cracked, there might be small crevices in the internal, which are not conducive to long-term collection, and will directly affect the aesthetic and economic value of the jade. Therefore, jade should not be put together with a hard object. When placing a jade article, ensure that it is steady. Besides, it is best to keep unused jade ornaments in jewelry bags or boxes.

- 白玉自在观音（现代）
  White Jade Avalokitesvara (*Guanyin*) Statue (Modern Times)

内部形成小裂痕，不利于玉石的长久收藏，会直接影响玉器的美观和经济价值。因此玉器不宜与硬物同置，摆放玉器时要确保平稳，不用时的玉饰最好是放进首饰袋或首饰盒内。

避免灰尘

净度是评定玉器的重要标准。出土的古玉肌理浸有污物，若再积灰遇污，会影响古玉恢复的时间。如果积尘，可用软毛刷、白布清洗。如果用有色的布擦玉，布的颜色有可能会染到玉上面，因而以干净、柔软的白布为好。

避免油污

油污会堵塞玉肌，不利于玉器的收藏。若有油污附着于玉器表面，可用温淡的肥皂水刷洗，再用清水洗净。对于受到严重污染的古玉可到专门清洗玉器的店里进行保养。

避免与化学物品接触

化学物品，包括香水、化学药剂、肥皂等会对玉器有侵蚀作用，会使玉器变色、变形、散发异味。

Avoid Dust

Clarity is an important criterion for evaluation of jade. Unearthed antique jades might have soaked dirt into its body, and if it gets stained again by dust, the time for it to revive into its original luster will be long. If there is dust, use a soft brush and a white cloth to clean it. If one uses a colored cloth to clean the jade, the color of the cloth might stain the jade;thus a clean and soft white cloth is the best.

Avoid Oil

Oil could stifle the skin of jade, which is not good for the collection. If the surface of a jade is stained by oil, one can clean it first with warm and light soap water, then rinse it with pure water. For the heavily polluted antique jade, one can bring it to a jade cleaning and treatment shop for maintenance.

Avoid Contact with Chemicals

Chemicals, including perfumes, chemical medicines, soap, etc., have a corrosive effect on jade. They can make jade change its color and shape, and even emanate a smell.

保持湿度

过于干燥的环境容易使玉内水分蒸发，从而损害玉的品质。所以为了保持玉的润泽，要保护玉内的水分，不宜将玉器存放于干燥环境之中。

避冰避冷

玉器遇冷，玉肌会紧闭，遇冰更会导致玉质开裂，所以收藏玉器要避开冰冷的环境。

避光避热避火

玉器受阳光直射过久，玉的分子体积会增大，玉质受到影响。存放环境过热，也对玉质不利。玉器遇火会变色，影响观赏性。

- 白玉灵芝大花插（清）
  White Jade Large Container for Cut Flower with Ganoderma Lucidum (*Lingzhi*) Design (Qing Dynasty, 1616-1911)

Maintain the Humidity

Too dry an environment makes it easy for the water inside the jade to evaporate, which is to the detriment of the quality of jade. Therefore, in order to maintain the moisture of jade, one should not store jade in a very dry environment.

Avoid Ice and Cold

When jade meets the cold, the pores on the jade skin will be closed, and when it meets ice, it might be cracked. Therefore, to avoid the cold and icy environment is very important to jade collection.

Avoid Light, Heat and Fire

If exposed to direct sunlight for too long, the molecular volume of jade will increase, and then the quality of jade is affected. An overheated storage environment is also detrimental to the quality of jade. When jade meets fire, its color will change, which affects the aesthetic value.